ENGLISH HERITAGE

ENGLISH HERITAGE RESEARCH TRANSACTIONS
RESEARCH AND CASE STUDIES IN ARCHITECTURAL CONSERVATION

METALS

*Volume **1***

April 1998

JAMES X JAMES

© Copyright 1998 English Heritage

All rights reserved. No part of this publication may be reproduced, stored in a retrieval system or transmitted in any form or by any means, electronic, mechanical photocopying, recording or otherwise, without the prior written permission of the copyright owner and the publisher.

First published by James & James (Science Publishers) Ltd, 35-37 William Road, London NW1 3ER, UK

A catalogue record for this book is available from the British Library
ISBN 1 873936 62 1
ISSN 1461 8613

Series editor: Jeanne Marie Teutonico, English Heritage
Consultant editor: Kate Macdonald

Printed in the UK by Redwood Books Ltd

Disclaimer
Unless otherwise stated, the conservation treatments and repair methodologies reported in the research papers and case studies of this volume are not intended as specifications for remedial work. English Heritage, its agents and publisher cannot be held responsible for any misuse or misapplication of information contained in this publication.
The inclusion of the name of any company, group or individual, or of any product or service in this publication should not be regarded as either a recommendation or endorsement by English Heritage or its agents.

Accuracy of information
While every effort has been made to ensure faithful reproduction of the original or amended text from authors in this volume, English Heritage and the publisher accept no responsibility for the accuracy of the data produced in or omitted from this publication.

Front cover
The lead covered roof of the church of St Cross, Winchester
(Bill Bordass/William Bordass Associates, London)

Contents

Acknowledgements *v*

Foreword
Sir Jocelyn Stevens CVO, Chairman of English Heritage vii

Preface
John Fidler RIBA ix

Introduction to the *Research Transactions* series:
John Fidler 1
Introduction to Volume One: *Metals*
Jeanne Marie Teutonico 6

Part I: Research
The behaviour of structural cast iron in fire:
A review of previous studies and guidance on achieving a balance between
improvements in fire protection and the conservation of historic structures
Alan Porter, Chris Wood, John Fidler and Iain McCaig 11

Underside corrosion of lead roofs and its prevention:
an overview of English Heritage's research
Bill Bordass 21

Sideflash in lightning protection: a report on tests with masonry
Norman Allen 73

Part II: Development
Case study: The application of cathodic protection to historic buildings:
buried metal cramp conservation in the Inigo Jones Gateway,
Chiswick House grounds, London
Keith Blackney and Bill Martin 83

Case study: The conservation of the lead sphinx statue at Chiswick House
Keith Blackney and Bill Martin 95

Case study: Development and long-term testing of methods to clean and coat
architectural wrought ironwork located in a marine environment: the
maintenance of railings at the Garrison Church, Portsmouth
Keith Blackney and Bill Martin 103

Annex: English Heritage's research programme: a schedule of projects
on historic building materials decay and their treatment 1992/3–1997/8 127

Acknowledgements

Thanks are due to the staff of English Heritage's Architectural Conservation team for their hard work in bringing this first volume of the new Research Transactions series to fruition. Transactions series editor and research programme coordinator, Jeanne Marie Teutonico, has taken overall responsibility for the programme; Bill Martin, Chris Wood and Keith Blackney acted as research project managers and contributed to the outputs of the research reported here. Other English Heritage staff and consultants were also involved in managing projects or formed part of the review process and here Keith Ranson and Jim Child (Mechanical & Electrical Engineering, Professional Services Division), David Heath (Principal Architect, Conservation Group) and former cathedral architects Harry Fairhurst FRIBA and Corinne Bennett ARIBA must be acknowledged for their constructive contributions to the work.

The consultant editor for Volume I was Dr Kate Macdonald to whom English Heritage is indebted for her assistance in bringing together and structuring the work of several diverse authors into a coherent and professionally produced publication. The organisation is also grateful to our anonymous, independent, external peer reviewers and members of our Cathedrals Research Client Liaison Committee (including Dr Richard Gem of the Cathedrals Fabric Commission for England and Julian Limentani RIBA, Hon Secretary of the Cathedral Architects Association and Surveyor of the Fabric of Peterborough Cathedral) for their efforts in the service of technical and scientific quality control.

English Heritage also thanks its consultants; Dr Bill Bordass (William Bordass Associates, London); Dr David Farrell and Kevin Davies (Rowan Technologies, Manchester); Dr Norman L Allen (UMIST, Manchester); Alan Porter (Warrington Fire Research Consultants, Croydon); George Ballard (GB Geotechnics, Cambridge); and Gareth Glass (formerly of Taywood Engineering, London) for their ingenious, methodical and innovative work.

Acknowledgement and grateful thanks for contributions to this research are also given to the following collaborating institutions and research organisations that have either shared our costs or provided additional leverage on our resources through aligned, complementary work. These include the Historic Royal Palaces Agency; the Lead Sheet Association; the National Trust; Liverpool John Moores University; the Scottish laboratory of the Building Research Establishment; the Department of Electrical Engineering & Electronics, University of Manchester Institute of Science & Technology; and the Interface Analysis Centre of the University of Bristol.

Foreword

I am delighted to welcome readers to English Heritage's *Research Transactions* series. This entirely new venture seeks to draw together and to make publicly available the very considerable body of scientific and technical knowledge that is being produced from our work in conserving the nation's historic buildings and monuments. For too long our chief assets – the brains and skills of a dedicated and highly experienced workforce – have lacked a credible corporate platform to exhibit their discoveries and conclusions in conservation for the public good. Now, at last, we can share our ideas and develop with our peers new standards of practice for the coming century.

It is hoped that the *Transactions* will make the primary findings from our research and case work more accessible to other scientists and testing engineers as well as to expert architects, structural engineers, surveyors, conservators and policy-makers. These are not textbooks for the novice or the layman. They will underpin the technical advice and published guidance that English Heritage gives to government, local authorities and to the public on conservation matters generally. They are a record of our current understanding: helping to develop policies and standards of practice at the cutting edge of conservation.

Besides this inaugural volume on *Metals*, another four volumes are currently being prepared for publication during the next twelve months. Two volumes will deal with a variety of subjects covering studies of the performance and treatment of porous building materials; one volume will be dedicated to a study of earthen construction; and the fourth volume brings together reports on certain aspects of timber decay and its treatment: outputs mostly from the first five years of strategic technical research generated by our cathedrals grants programme. Further volumes will follow as the growing programme of research gathers momentum and new projects are concluded.

English Heritage is one of the few conservation bodies world-wide that is in the fortunate and highly influential position of being not only able to develop applied research needs from its own casework but also to collaborate with others in the delivery of practical research results, and the interpretation and dissemination of the findings through policy, guidance and practice to those concerned with conservation in the field.

I commend the series and this volume to an informed international audience and look forward to the scientific feedback and dialogue it is intended to stimulate.

Sir Jocelyn Stevens CVO
Chairman of English Heritage
March 1998

Preface

Metals, the launch publication of our new *Research Transactions* series, is likely to bring a new and discerning audience to English Heritage. It may include the scientific and test engineering communities, perhaps unfamiliar with the ethical concepts and technical subtleties of heritage conservation; surveyors, specifiers and conservators unused to, or unpractised in, the physics, chemistry and biology that lie behind their art; decision-makers in the public sector, industry and commerce, looking to sharpen their technical policies or enter new markets; and students and academics in pursuit of interesting ideas and knowledge. Each group will have its own perspective of, and area of interest in, the subjects to be covered in this set of volumes but it falls to me to set out our context for the organisation's contribution to the field.

For those unfamiliar with our work, I should restate the fact that English Heritage is the principal organisation statutorily responsible for the conservation of the historic built environment in England. It was established as a non-departmental public body by the National Heritage Act 1983 and began work in April the following year. We maintain four hundred historic sites and open them to the public. We are the government's official advisors on all matters concerning the conservation of the built heritage and are a major source of public funding and the national centre of technical expertise for archaeology and the repair of conservation areas, listed buildings and scheduled ancient monuments.

Among our chief duties, we are charged to *secure the preservation* of historic buildings and ancient monuments and to *advance knowledge* of their preservation. But these are very tall orders indeed! The potentially damaging impacts of global warming, atmospheric pollution and mass tourism upon ageing historic buildings, for example, are now compounded by shortages of traditional repair materials and skill gaps in the building industry. Within current resources, our mission then must be to maximize our influence and harness collective efforts to pass on the country's architectural legacy to the future in a healthy condition.

But the technical aspect of building conservation is a relatively recent and slowly developing field in which the internationally agreed body of knowledge on historic material decay processes and treatment systems is small, experimental and often heavily reliant upon what are now considered to be rudimentary technical perceptions and folk traditions. There is a pressing need for systematic technical and scientific inquiry to increase the sum of knowledge: to make regimes of preservation and/or repair more benign and cost-effective for the future.

Well-meaning though ill-informed repairs can be just as harmful as wilful neglect. If we can better understand the ways in which ancient stones or metal decay, and find ways to accurately monitor their rates of deterioration, we can make more informed judgements about how and when it is best to intervene to save them. Treatments then can be fine-tuned to suit particular circumstances: early preventive care and regular maintenance may suffice, or works can be postponed until the limit of the component's survival is reached; at which point remedies can be closely targeted to minimize effects on only the most severely damaged parts.

Research can also tell us when refurbishment fails to work when treatments only address the symptoms and not the causes of decay. Experiments in transferring technology from, say, medicine or marine engineering, may resolve long-standing problems of remote sensing or ways to minimize the impact of repairs. Key-hole surgery has left the hospital and come to building sites!

The testing of new materials and repair techniques provides us with an insight into the long-term implications of the multiple retreatment of historic fabric – we are not the first and we shall not be the last generation to care for historic buildings, and the cumulative effects of renovation are only now beginning to be recognised. I am proud to say that English Heritage has been at the forefront of this conceptual leap and instrumental in devising long-term proving trials in the field, when others have stayed in the laboratory.

Research informs what we do and helps us to do it better. Long-held assumptions about the character and performance of traditional repair materials and of treatments, that have never been objectively assessed before, can now start to be challenged and remedial practice standards revised to make improvements. Of course, this can lead to friction – old ways are slow to change. But we are confident that we can carry people with us by force of argument to shift technical habits in building conservation to a more benign, practical and cost-effective way of working.

Our work is not 'blue sky' research carried out in the abstract by white-coated scientists in their ivory towers – it is *applied* research and development based upon real-life problems for which hard-nosed, *practical* solutions are demanded. Our design, commissioning and management team for the research is multidisciplinary, involving architects, surveyors and conservators who know what is required from the specialist skills of their scientific collaborators, consultants and contractors. The organisation places great emphasis upon keeping end-users involved in the research process. By articulating the conservation problems and by steering the research, our practitioners can help the scientists to find ethically and technically appropriate solutions.

Of course, one of the great frustrations in research management is not always being able to deliver pertinent new information in time for end-users to apply. Readers will understand that building materials research using proper scientific methods simply cannot be rushed. Materials placed in an accelerated weathering chamber, for example, need first to be prepared, cured, weighed and measured on calibrated machinery and the processes repeated many times if the results are to have any kind of statistical accuracy. And even then, artificial chambers never exactly replicate external conditions of exposure. The research teams working for English Heritage are as keen as anyone to broadcast their findings as quickly as possible whether the results be good or bad. But scientists are necessarily concerned to check the quality and standard of results first because of the high-profile nature of the work and because of the ethical, technical and economic impact any policy recommendations will have subsequent to the research.

These are important steps that sometimes make part of English Heritage's work seem slow. But the welfare of our historic building stock in the next century and beyond is important enough for us to make sure we get things right. We are, of course, concerned to get value for money from research. But we are equally careful to base our technical advice and conditions of grant support for the care of historic buildings on the best possible scientific evidence.

The perceived delays in delivering practical results can sometimes be frustrating for the specifiers and conservators who desperately need the latest state-of-the-art advice to save monuments in the most sensitive and sensible way. We heard of one case, for example, where a keen conservation officer in a local planning authority made it a condition of listed building consent that an architect specify mortar mixes for the repair of a Grade II listed building, 'in accordance with the findings of English Heritage's technical research'. But our work on mortars is not yet conclusive or finished. At least one more phase of testing is required to complete the work and provide useful guidance.

The research managers in our Architectural Conservation team, together with their consultants and contractors, can be contacted via the *Transactions* editor to discuss any fundamental scientific and technical matters arising from individual research projects published here. Specific technical advice, however, to local authorities and to historic building owners and their professional advisors, based on the research, remains the responsibility of the regional teams in English Heritage. Their technical policy stance regarding material recipes and practice remains unchanged until research findings are ultimately debated and corporate changes to advice are formally announced as, for example, technical policy statements.

To conclude: we have addressed long-standing, common technical problems about the decay and treatment of fragile historic building materials, from a conservative ethical position which follows the tenets of the Venice Charter (ICOMOS, Paris 1966), which states that interventions should only take place where they are urgently necessary, minimal, potentially reversible and non-prejudicial to future interventions, in order to protect the archaeological integrity and special architectural or historic interest of our cultural patrimony. Our long tradition and practical experience of repairing buildings and monuments, and our vision of future improvements in the technical practice of conservation, have now harnessed imaginative science, technical flair and practical ingenuity to push forward the boundaries of knowledge and understanding in the field.

Although the built heritage makes up less than 4% of the United Kingdom's total building stock, the discoveries made here have a much wider application and will surely help to inform general maintenance practice. We hope that the dissemination of the results of our research and development work, through publication in the following pages of scientific reports and case studies, and future volumes of our *Transactions* series, will enhance general practice and lead to more focused, benign and cost-effective care for buildings of every age, type and location.

John Fidler RIBA
Head of Architectural Conservation
English Heritage

Introduction to the *Research Transactions* series
Historic building materials research at English Heritage

JOHN FIDLER
English Heritage, 23 Savile Row, London W1X 1AB

BACKGROUND

Research into the decay and treatment of historic building materials in England is the province of the Architectural Conservation Team within English Heritage.[1] Through aligned and collaborative ventures, including research sponsored by the European Commission and agreements and contracts with over 15 national and international groups of consultants and contractors, it indirectly employs more than 30 specialists on 23 projects in the service of better building conservation.

Although research work is modest in scale by comparison with the organisation's main activities, and currently constitutes less than 0.2% of total annual expenditure, it provides a much needed service that underpins English Heritage's reputation as a centre of excellence. It is a national focus for advancing knowledge, standards and skills in the technical and scientific aspects of building conservation, in the study of the decay of historic building materials and in the development of benign, practical and cost-effective treatments.

This modest though important programme will repay itself many times in the future through savings made, on the basis of the results, on more appropriate, targeted and efficient treatments for decaying historic building materials. English Heritage has generated significant gearing on its investment by working with other researchers and hopes to increase its leverage in the future through extended partnerships and more collaborative research. But how did an English Heritage research programme come into being at all?

HISTORIC PRECEDENTS

The story of state-sponsored scientific enquiry into the decay and treatment of building materials in the United Kingdom has a long and fascinating history. Limitations of space preclude a detailed examination of the subject here but some edited highlights should provide the reader with an idea of progress made over the last two hundred years. For the most part, with one or two exceptions, development of the subject can be characterized by sporadic, disjointed efforts on the part of individuals who, through their own vision and talent, were able to stimulate either direct scientific action themselves or such interest in others. Where significant scientific testing programmes were established, their connections to building conservation have usually been indirect and informal, though nonetheless important and influential.

In the early nineteenth century, for example, the first treatise on the timber decay fungi dry rot (*Serpula lacrymans*), was written for the Navy Board.[2] Although it specifically focused on ships' timbers, the report's influence and that from similar contemporary studies was felt throughout the construction industry with timber preservative treatments being developed and patented in response throughout the century.

In 1839, a report to the Commissioners of Woods and Forests (who were then in charge of public building) recommended the choice of stone for the construction of the new Houses of Parliament and caused the bringing together of a national collection of building stones.[3] A systematic scientific approach to choosing materials based upon their characteristics and relative performance had not until then ever been commissioned by the British government and it received much praise at the time and later, despite the consequent failings of the recommended materials.[4] In fact, as soon as the Palace of Westminster was completed in 1861, another committee was appointed by the First Commissioner of Public Works and Buildings to investigate the exceedingly rapid decline of the chosen stone and it sponsored accelerated weathering tests to assess the stone's comparative durability and look at possible stone preservatives.[5]

In the same period in the private sector, the architect George Gilbert Scott had been carrying out his own trials with preservatives on decaying stone at Westminster Abbey.[6] His interest and state concern combined when he was commissioned to restore the Chapter House there as a secular visitor attraction, after the public records stored in the building were moved to Chancery Lane in 1868. Scott's excellent restoration of the building was later to incorporate experiments with Baryta water (barium hydroxide) used as a stone consolidant.[7] The unsuccessful results were published by the government in 1904 (Church 1904).

Public sector research into the conservation of historic materials only really emerged after the Ancient Monument Amendment Acts of 1900 and 1910, when the Office and Ministry of Works was given technical responsibility for the care of state monuments. In 1912, Frank Baines became Chief Architect for Ancient Monuments and

Historic Buildings and started regularizing conservation practice (Saunders 1983; Chitty 1987). He also formed a close working relationship with government scientists, a tradition that continues to this day. His published revisitation of the performance trials of the stone consolidants at the Chapter House of Westminster Abbey are among the first in the field (Office of Works and Public Building 1914).

Between the World Wars, the Department of Scientific and Industrial Research[8] carried forward complementary programmes to understand the characteristics and performance of a range of historic building materials, including seminal reports on lime and lime mortars, terracotta and stone (Cowper 1927; McIntyre 1929; Schaffer 1932 respectively) which still have great relevance today. The work, based upon careful field observations, international literature reviews and laboratory testing, brought British scientific understanding of the period to an international high point that has been hard to surpass.

In the 1950s the now famous Ancient Monuments Laboratory started to emerge from the archaeological finds workshop of the Ministry of Works and occasionally dabbled in the scientific assessment of decaying building materials as well as the cleaning, care and analysis of moveable artefacts destined for museums and display (Biek 1963). A very wide range of materials seems to have taxed the Laboratory, including the decay of fourteenth-century timber piles, the protection of cast-iron roof plates at the Palace of Westminster, gamma radiography of fallen stones from Stonehenge and the analysis and matching of Roman *opus signinum* mortars from various archaeological sites. The investigations were a collaborative effort between architects, archaeologists, conservators and the in-house chemists within the conservation team and specialist scientists in various government departments including the laboratory of the Government Chemist, the Paint Research Station and the Atomic Energy Authority at Harwell.

By 1968, the Ancient Monuments Laboratory, now in the Directorate of Ancient Monuments and Historic Buildings (DAMHB, Ministry of Housing and Local Government), was concentrating much more on the study and conservation of archaeological finds and artefacts because of the growing demand from that sector. Material science continued to play a significant role in the service of building conservation through the creation of a Special Works Architect post by the then Superintending Architect, Patrick Faulkner, to focus technical inquiries within the organisation and to liaise with external scientific support, the precursor of our research procurement teams today. Work of the period included, for example, field trials of surface coatings and shelter coats for friable stone at Audley End.

In the 1970s, the post's responsibilities and demands for services grew so much that a small branch was formed, called the Research and Technical Advisory Service (now in DAMHB, Department of the Environment). Its modest, informal, though highly creative, scientific collaboration with, and influence upon, scientists at the Building Research Establishment during the early 1980s led to a spate of influential publications (see Ashurst & Clarke and BRE). By the end of the decade much of the established wisdom from this and other research, and from experience of practical problems in the field, enabled English Heritage to publish its first five volumes of technical handbooks, *Practical Building Conservation* (Ashurst & Ashurst 1988/9) which have become seminal textbooks on the subject and international bestsellers.

In 1991 a change of management and direction at English Heritage was partly stimulated by the injection of new resources, when the government of the day provided new additional funds to give as grants to English cathedrals for their repair and conservation. It was argued that many of these venerable buildings possessed similar problems of decay and deterioration and that a strategic technical research programme should be set up in parallel with the grants scheme to inform technical and scientific decisions about remedial work. For the first time, the state service for building conservation was empowered with sufficient resources to commission large-scale, formal research projects. The story is up to date.

THE CURRENT RESEARCH PROGRAMME & ITS MANAGEMENT

English Heritage's strategic technical research programme was developed in response to the research needs articulated by the survey of English cathedrals carried out by Harry Fairhurst FRIBA, Surveyor Emeritus of the Fabric of Manchester Cathedral and consultant to English Heritage, prior to the establishment of the organisation's grants scheme. Common technical and scientific problems were identified and addressed in the design of the research projects. Research themes, as set out in *English Heritage's Research Programme* in the Annex to this volume, included the characterization of materials and their decay mechanisms; short- and long-term evaluations of treatment regimes and their repeated use; and risk assessment strategies for deciding on when and how to intervene. Testing techniques thus developed for one project could be utilised on other projects to save time and money.

Of course, in a perfect world, it would have been desirable to carry out scientific testing in advance of any repair programmes. But the financial constraints of the grant scheme and the urgent needs of the cathedral authorities limited this option in favour of a more pragmatic approach, where the technical and scientific investigations might benefit from the immediacy of the refurbishment projects and vice versa. Needless to say, many of the common building faults identified in the cathedrals for scientific enquiry were also applicable to a very much larger constituency of historic buildings and their interests.

A Cathedrals Research Client Liaison Group was set up to monitor and steer the research with the programme's project managers. It consists today of senior technical and managerial staff within English Heritage and their consultants who are involved in providing technical advice and/or grant aid to cathedrals (two of them are former cathedral architects). Independent external members include the Secretary of the Cathedrals Fabric Commission for England (who has a technical policy interest in the research findings), and a practising cathedral architect (who is also a member of English Heritage's Cathedrals and Churches Advisory Committee, Secretary of the Cathedral Architects Association and a long-standing committee member of the Ecclesiastical Architects and Surveyors Association).

Scientific peer reviews of the research, to test for scientific quality, clarity of reporting and usefulness, are undertaken by the Liaison Group and at an informal technical level by cross-referring reports to other contractor teams within the programme where complementary studies are taking place. External high-level peer reviews are also undertaken for complicated time-consuming reports through submissions to international conferences as papers for publication in proceedings or to learned journals both in the UK and abroad. The project managers also report on the technical and scientific content of their work to English Heritage's Science and Conservation Panel (which advises its Commissioners) and account for the work annually to the organisation's Cathedrals and Churches Advisory Committee. The Architectural Conservation team also takes soundings from the end-user groups within English Heritage, from the professional meetings of its staff architects, historic area advisors and inspectors and by consultations with industry and trade associations, where appropriate.

Marks of success in terms of the quality of research include the award of the Torrey Fuller Award by the North American Association for Preservation Technology for the publication of project AC1, *The Smeaton Project* (Teutonico et al 1994); the establishment of a patent by research consultants on English Heritage's behalf for a innovative, benign yet simple treatment to safeguard against underside lead corrosion; and the published acknowledgement of consultation with three of English Heritage's research project managers in a current international review of stone conservation research.[9]

Most of the research projects are of between two to five years' duration and involve a substantial amount of time spent on literature surveys to encompass the international state of knowledge and trends in the subject area. Further effort is spent on developing the sometimes painstakingly slow protocols often required to establish detailed scientific facts. Programmes of monitoring and testing can then follow to accumulate pertinent scientific data. For example, field assessment and monitoring of deterioration processes often entail several years of observation and measurement to account for seasonal variations in micro-climates that may instigate the real cause rather than the current symptoms of decay. Laboratory analysis and experimentation complement this work before site trials take place so that data can be synthesized, written up and interpreted before conclusions are made known to the widest possible audience.

Subsequently, programmes of publication are formulated and implemented. At the highest level, complex scientific papers are published by their originators in recognised nationally- or internationally-reviewed journals among the contractors' peers. Simplified and interim research reports are made available through papers delivered at conferences proceedings (Ashall et al 1996) or through the English Heritage *Research Transactions* series. Developments from the research in the form of technical policy statements, technical guidelines and advisory notes are then disseminated at a more basic level.[10]

English Heritage's stated duties in the National Heritage Act require it to secure the preservation of the historic built environment and to advance the public's knowledge of ancient monuments, historic buildings and their preservation. Its functions allow it to give advice (of any type including scientific and technical advice) and 'carry out, or defray or contribute towards the cost of research' (HM Government 1983, Section 33 [2] [c], 20). The terminology of the Act differentiates between these last-mentioned functional verbs and the essentially reactive process of grant aid, referred to elsewhere in the document, so the organisation does not run any open-access research grant programmes nor does it advertise targeted calls-for-proposals, much to the frustration of some in the national research community. We do not have powers to act in these ways.

So legal constraints dictate that English Heritage operates research programmes through direct tendered procurement with its own and other acquired funds. It also collaborates with others, in the UK and overseas, in aligned projects, either bidding for grants or operating on a shared cost basis. The type and quality of the research therefore can be contrasted with the more open structure of directed or open access grant programmes, such as those of the Department of the Environment, Transport and the Regions (Partners in Technology) or of the European Commission Directorate General XII (Environment Framework). Bidding for external funds can, of course, potentially force English Heritage to shift the focus of its research priorities to the grant giver's agenda, though we guard against this and search for common interests and synergy wherever we can.

THE PUBLICATIONS PROGRAMME

The results of past, current and future research work will be worthless unless they are widely broadcast to fellow scientists, engineers and technicians and the conservation community of building owners, specifiers, quality controllers, conservators and craftsmen. It is our intention to publish final reports of testing results as soon as is reasonably practical, given the limitations on our resources, and to promulgate policies and guidance based on our findings thereafter.

Communication is often very limited between scientists and testing engineers on the one hand and the end-users of research, the architects, surveyors, structural engineers and conservators, on the other. Work is often done in a fragmentary way on single monuments or buildings without enlarging the scope to wider applications and without adequate exchange of ideas and experience across borders (Macioti 1991). English Heritage's

Research Transactions series will set out our efforts to describe and explain scientific and technical questions concerning historic building materials and their welfare for all to see, hopefully prompting and contributing to a dialogue for mutual benefit and concern.

Internally-produced publications always run the risk of being criticised as so-called 'grey literature' (Benarie 1992). The technical and scientific papers that here follow have all been internally and externally peer-reviewed from two separate, though complementary perspectives. From the scientific community, we wanted to know whether the science was good, the results made sense and the papers accurately described the processes and conclusions involved. From the conservation community of practitioners who will now hopefully make use of the research in their work, we wanted to know if the research was understandable, logical and had actually answered the original queries and problems upon which the work was based. Only informed readers can ultimately judge if we have succeeded.

Our aim is to encourage better co-operation; to prompt interest in and act as a catalyst for further technical and scientific investigations and developments in the United Kingdom and in other scientific and conservation communities around the world; to assist in the current international review of knowledge; and to obtain feedback and ideas to help define future lines of national research for the benefit of our cultural patrimony.

ENDNOTES

1 Broad international equivalencies for the study of historic building materials in conservation science can now be found in many developed countries. For example, in France, the *Architectes en Chef des Monuments Historiques* have direct access to several laboratories and scientific establishments; in Italy, there are the fine laboratories of the Instituto Centrale di Restauro and of the Consiglio Nazionale delle Ricerche in Rome and others at the International Centre for the Study of the Preservation and Restoration of Cultural Property (ICCROM); in the United States, among others, materials conservation science can be found at the Getty Conservation Institute in Los Angeles and at the Architectural Conservation Laboratory at the University of Pennsylvania.

2 See Bowden 1815 (information kindly provided by Brian Ridout of Ridout Associates, Stourbridge UK, to be expanded in the forthcoming book from English Heritage and Historic Scotland, *Timber decay and its treatment: the conservation approach*).

3 See The Commissioners for the Stone for the New Houses of Parliament 1839. See also Papworth's comments in his *Gwilt's Encyclopedia of Architecture* (Papworth 1888, 452, para 1663). The stone collection now forms the main part of the reference material in the geological section of the Natural History Museum, London.

4 Pugin's specification was deployed between 1840 and 1860, principally in Bolsover and Anston magnesian limestones that did not perform well in London's notoriously sulphurous atmosphere. The committee was appointed on 23rd March 1861 'to inquire into the decay of the stone of the new Palace of Westminster and into the best means of preserving the stone from further injury'. Its response was printed by Parliament in 1861 (Crookes 1861).

5 For example: (1832) J Kayan's corrosive sublimate of bichloride of mercury; (1836) Sir William Burnett's zinc chloride; (1837) M Breant's iron sulphate; (1838) Bethell's injection of *tar oil*; (1842) V Payne's two part system of iron sulphate and sodium carbonate: all reported in *Gwilt's Encyclopedia of Architecture* (Papworth 1888, 507, para 1752).

6 Scott was appointed to the post of Surveyor of the Fabric in 1849 and carried out stone conservation experiments with shellac and spirits of wine on various internal decaying areas, apparently, he reported, with some success (Scott 1879 [1995], 153, fn 5). Externally, the treatment did not perform well except in sheltered areas and included part of the cloisters close to the entrance to the Chapter House.

7 The Chapter House was originally built for ecclesiastical and secular functions, acting as both the Crown treasury and the seat of governance for the abbey. Some of the first parliaments were held there and it has remained in Crown rather than Church hands to this day. For several hundred years it functioned as a records store and was restored by Scott between 1861-8. The stone preservative treatment applied post-dates Scott's work but used a barium hydroxide solution devised by A H Church.

8 Later to become the Building Research Station, then the Building Research Establishment, an agency of the Department of the Environment, and now BRE Ltd, a private company owned since March 1997 by the Foundation for the Built Environment, a non-profit distributing body consisting of 125 member organisations drawn from a wide spectrum of construction interests. The BRE's main site and library remains at Garston, Watford, Hertfordshire WD2 7JR; tel: 01923 664000, fax: 01923 664010, email: enquiries@bre.co.uk. See also the Internet page at: www.bre.co.uk.

9 See the acknowledgements to English Heritage's Architectural Conservation staff, authors in this Transactions volume, in the preface to Price, *Stone Conservation: an overview of current research* (1996), page xi.

10 For example, Technical Policy Statement No 1, *Hybrid mortar mixes containing a blend of non-hydraulic lime and hydraulic lime binders* (English Heritage Product Code XH20061, London, May 1997) warns of technical problems when specifying particular mortar recipes and places a moratorium on their use pending further planned research. *The use of intumescent products in historic buildings* (English Heritage Product Code XH20055, London, May 1997) gives guidance on particular passive fire safety measures. *Lead roofs on historic buildings* is an advisory note on underside corrosion published jointly by English Heritage and the Lead Sheet Association (English Heritage Product Code XH20055, London, April 1997). These texts were all developed from the organisation's applied research programme.

BIBLIOGRAPHY

G. Ashall, R. Butlin, J. M. Teutonico & W. Martin, 'Development of Lime Mortar Formulations for Use In Historic Buildings: a report on Phase II of the Smeaton Project, a joint research programme of English Heritage, ICCROM, Bournemouth University and the Building Research Establishment', in *Proceedings of the 7th International Conference on the Durability of Building Materials and Components, 19-23 May 1996, Stockholm, Sweden*, ed C. Sjostrom (London: Spon, 1996), 352-60.

J. Ashurst & B. L. Clarke, *Stone Preservation Experiments*, Building Research Station Report (Watford: BRE, 1972).

J. Ashurst & N. Ashurst, *Practical Building Conservation*, English Heritage Technical Handbooks. **1**, Stone Masonry; **2**, Brick,

Terracotta and Earth; **3**, Plasters, Mortars & Renders; **4**, Metals; **5**, Wood, Glass and Resins (and consolidated bibliography) (Aldershot: Gower Technical Press, 1988/9).

M. Benarie, ' The Greying of Literature', in *European Cultural Heritage Newsletter*, **6** (5) (Brussels: European Commission DG XII for Science, Research and Development, December 1992), 3-5.

L. Biek, *Archaeology and the Microscope: the scientific examination of archaeological evidence* (London: Lutterworth Press, 1963).

A. Bowden, *A Treatise on the Dry Rot* (London: Navy Office, 1815).

Building Research Establishment, *Colourless Treatments for Masonry*, BRE Digest 125 (Watford: BRE, 1971).

Building Research Establishment, *Decay and Conservation of Stone Masonry*, BRE Digest 177 (Watford: BRE, 1975).

Building Research Establishment, *The Selection of Natural Building Stone*, BRE Digest 269 (Watford: BRE, 1983).

Building Research Establishment, *Cleaning External Surfaces of Buildings* BRE Digest 280 (Watford: BRE, 1985).

Building Research Establishment, *Control of Lichens, Moulds and Similar Growths*, BRE Digest 139 (Watford: BRE, 1989).

G. Chitty, A Prospect of Ruins, *Transactions of the Association for Studies in the Conservation of Historic Buildings*, **12** (Kilmersden: ASCHB, 1987), 43-60.

A. H. Church, *Copy of memoranda by Professor Church FRS, furnished to the First Commissioner of His Majesty's Works, etc., concerning the treatment of decayed stonework in the Chapter House, Westminster Abbey*, Parliamentary Command paper 1889 (London: HMSO, 1904).

A. H. Church, *Improvements in the means of preserving stone, brick, slate, wood, cement, stucco, plaster, whitewash, and colour wash from the injurious action of atmospheric and other influences, etc*, British Patent 220, 28th January 1962.

Commissioners for the Stone for the New Houses of Parliament, *Report of the Commissioners appointed to visit the quarries and to inquire into the qualities of the stone to be used in building the new Houses of Parliament* (London: HMSO, 1839).

A. D. Cowper, *Lime and Lime Mortars* Building Research Special Report **9** (London: HMSO, 1927).

W. Crookes, *Report of the Committee on the Decay of the Stone of the New Palace at Westminster*, Parliamentary Command Paper, (London: HMSO, 1st August 1861).

English Heritage & Historic Scotland, *Timber Decay and its Treatment: the conservation approach* (London: E & FN Spon, forthcoming).

Her Majesty's Government, *The National Heritage Act 1983* (London: HMSO, 1983).

M. Macioti, 'Science, Technology and European Cultural Heritage, in *Proceedings of the European Symposium on Science, Technology and European Cultural Heritage. Bologna, Italy. June 1989*, eds N. S. Baer, C. Sabbioni & A. I. Sors (Oxford: Butterworth-Heinemann for EC DGXII, 1991).

W. A. McIntyre, *Investigation into the Durability of Architectural Terracotta and Faience* Building Research Special Report **12** (London: HMSO, 1929).

Office of Works and Public Building, *Report of the Inspector of Ancient Monuments for the Year ending 31 March 1913*, Parliamentary Command paper **7258** (London: HMSO, 1914).

W. Papworth, *Gwilt's Encyclopedia of Architecture*, 9th edition (London: Longmans, 1888).

C. Price, *Stone Conservation: an overview of current research*. Research in Conservation series (Santa Monica: The Getty Conservation Institute, 1996).

A. D. Saunders, 'A Century of Ancient Monuments Legislation 1882 -1982', in *Antiquaries Journal*, **LXIII** (I) (1983), 18 - 33.

R. J. Schaffer, *The Weathering of Natural Building Stones* Building Research Special Report **18** (London: HMSO, 1932, reprinted Watford: BRE, 1972).

G. G. Scott, *Personal and Professional Recollections* (1879), ed G. Stamp (Stamford: Paul Watkins, 1995).

J. M. Teutonico, I. McCaig, C. Burns & J. Ashurst 'The Smeaton Project: factors affecting the properties of lime-based mortars', in *Bulletin of the Association for Preservation Technology*, **25** (September 1994), (3 & 4), 32-49.

Introduction to Volume I: *Metals*

JEANNE MARIE TEUTONICO
English Heritage, 23 Savile Row, London W1X 1AB

It may be helpful to provide a context and introduction to the research papers that follow in this volume of the *Research Transactions*. Based on the theme of architectural metals and their decay or deterioration problems, in various environments, the papers describe minimal, reversible and non-prejudicial interventions from a complete understanding of the processes involved. All of the papers demonstrate the fundamental objectivity of our research, refusing to rely on anecdotal or standardized responses to the specialized problems of historic fabric. Risk assessment strategies play an increasing role in our work as does the transferral of technology from other industrial applications. Recognising that treatments can be as damaging as illnesses for friable old materials, our response has been to tread carefully and as little as possible on ancient ground.

THE BEHAVIOUR OF STRUCTURAL CAST IRON IN FIRE

For the last thirty years, increasing efforts have been made in the United Kingdom by the conservation community to retain, adapt and re-employ redundant eighteenth- and nineteenth-century warehouses, mills and factories. Many of these buildings were originally constructed using cast iron for their principal structural elements. While the reuse of industrial buildings has now become fashionable and even profitable, there are still technical problems confronting owners, specifiers and regulators, concerning fire protection standards, because of a lack of understanding of cast iron and its performance in fire.

English Heritage's research continues the pioneering interest in the behaviour of historic structural cast iron of the former Greater London Council. Tests in the 1980s by its Scientific Branch and others in co-operation with the Council's Historic Buildings Division (which has formed part of English Heritage since the GLC's abolition) were never widely published. The paper presented here makes good that problem and gives the work both a fire safety and planning context.

The old-fashioned, stock response to cast-iron structures was to view them all as potentially dangerous in fire situations. Mythical tales of exploding columns and cracking beams seem to have entered the national engineering and fire-fighting psyche. But the GLC's work proved that a more subtle understanding of the behaviour of cast iron in fires was possible and could be readily calculated.

The fire engineering approach to fire safety in historic buildings now allows us to make better risk assessments without over-reliance on passive fire protection measures that can be visually damaging to the special architectural or historic interest of listed structures.

UNDERSIDE CORROSION OF LEAD ROOFS AND ITS PREVENTION

Lead sheet roofing has a long and illustrious history in the story of English architecture and it continues to make a unique contribution to the character and appearance of buildings, old and new, throughout our towns and cities as we approach the new millennium.

English Heritage has invested a great deal of public money in the conservation, repair and replacement of historic roofs clad in lead sheeting. Through work on our own properties and through grant aid to others, we have always sought first to extend the life of existing, original material. But where the damage is great and unsuitable historic fixings have contributed to the deterioration, we have agreed to the replacement of old roofs in new leadwork: confident of the high quality, durability and value for money that the material affords.

For a number of years, there has been a steady trickle of requests for technical advice, to ourselves and the Lead Sheet Association, about corrosion taking place on the underside of lead sheets, some of it quite soon after installation. Inevitably perhaps, concern has been expressed in some quarters about the wisdom of continuing to invest in the material. But the number of cases involving this phenomenon are very small and are far outweighed by the usual inquiries about problems concerned with the poor design, specification and installation of fixings for cladding.

Nevertheless, English Heritage decided to co-ordinate a technical response to this supposedly new phenomenon and held a colloquium on the subject which ultimately resulted in a research consortium being formed to tackle the problem. Besides the lead industry, we were joined by the Historic Royal Palaces Agency and the National Trust in an aligned programme of technical investigation and research involving several universities, the Building Research Establishment's Scottish laboratory and English

Heritage's consultants in environmental and corrosion physics.

Although the same problems are experienced with other types of metal cladding (examples of corrosion are sometimes worse), our study was limited to research into leadwork as this is by far the most widely-used metal roof covering on historic property.

The results of the study are presented here in an overview of underside lead corrosion. First of all, this form of deterioration is not at all a new discovery: paint-makers are documented as having 'farmed' roofs for the corrosion product, white lead, as far back as the eighteenth century. We find that there is no evidence to prove that changes of alloy content in modern lead sheet production have made the material more susceptible to corrosion.

Instead, a complex and inter-connected series of circumstances appear to combine to foster decay. They include changes in the way that some historic buildings are used today (eg more humid environments and intermittent heating), combined with the just-in-time production of lead sheet (where the process of protective patination has less time to develop between the rolling mill or foundry and the building site) and the peculiarities of individual construction, timing of installation and types of substrates involved.

Outputs from the research include a free technical guidance document for building owners, specifiers and contractors published by English Heritage and the Lead Sheet Association entitled *Lead roofs on historic buildings: an advisory note on underside corrosion* (English Heritage Product Code XH 20055); advice on methods of decay risk assessment and ways to limit the chances of the corrosion taking hold; an application for a patent for a simple method of remedial or preventive treatment; and, coming soon, the revision of the published technical standard on lead sheet roofing by the Lead Sheet Association.

Armed with the findings of this research, nervous specifiers should sleep more easily. We continue to have confidence in lead sheet coverings and stress the importance of careful design and installation as with all conservation work.

SIDEFLASH IN LIGHTNING PROTECTION

English Heritage has been concerned for some time about the intrusive nature of lightning conductors installed upon historic buildings and whether alternative options exist. Installations are generally carried out to the requirements of the British Standard Code of Practice 6651: 1992 and usually result in the establishment of a Faraday cage which can be visually unpleasant if not well-located and physically disruptive to the historic fabric in terms of fixings. Studies then have concentrated on the development of a modified risk assessment method so that a balance can be fairly struck between the actual needs for partial or full protection to safeguard property and life, and the proper conservation of the historic fabric.

The British Standard details a method of risk assessment based on the probability of a lightning strike, adjusted by a number of weighting factors. But closer examination of the weightings shows that they may not all be particularly relevant to historic buildings and, in such circumstances, the consequences of a strike may be more acceptable (subject to certain precautions), in conservation terms, than intrusive lightning protection.

According to the Ecclesiastical Insurance Group, the risks, for example, of injury or death from lightning when in or near an historic building are extremely small. There are far higher risks when walking in the park or playing golf. It is noted that lightning strikes to churches rarely cause fires. So the major risks to property fall into three categories: structural damage caused by a direct strike of an unprotected feature such as a church pinnacle; damage to electronic equipment caused by power surges; and *sideflash* of lightning to earthed metalwork such as service cables and pipework: the subject of research presented here.

Our research has sought to better understand the subtlety in the issues involved in sideflash situations so as to improve risk assessment capabilities. We now know that the assumptions made in the current British Standard for determining flashover voltage as a function of spacing may need to be amended. But the impact on historic buildings protection will vary: in some circumstances, more earth bonding of metalwork will be required and its installation must be carefully handled to minimize its impact, but in the majority of situations, the distances between unprotected material, if correctly calculated, will be sufficient and no changes need be made.

THE APPLICATION OF CATHODIC PROTECTION TO HISTORIC BUILDINGS

Metal fixings in mass masonry walls have been employed in building since the time of the ancient Greeks and the Romans brought the technology to the British Isles. Their rediscovery here during the Renaissance and general adoption over the last three hundred years, especially for dressed ashlar walling, has been widespread and sometimes misunderstood.

Christopher Wren specified the protection of buried wrought-iron cramps from rusting by their encapsulation in molten lead, set back from the rain-soaked face of his masonry in dry sockets deep in the stonework. But others were less discerning and a common form of stone degradation found today involves the discolouration of stonework and its bursting apart through the action of corroding metalwork inside.

Iron can expand up to seven or eight times its volume in the process of corroding by the action of water and oxygen on the metal. This swelling of cramps causes enormous pressures on the surrounding masonry which ultimately fails in shear, leaving gaping brown stained cavities and pock-marked elevations.

The traditional method of repair has been to wait until this process occurs and then to cut out the offending damaged stone and metalwork, sometimes replacing the latter with a non-ferrous equivalent where structural stability is necessary. The pockmarks are then either patched over in lime-based mortars to match the surrounding stonework, or are filled in with new pieces of stone to match, in a process sometimes called indenting. But the problem then is that the stone facade starts to take on a changed appearance as indented stone rarely follows the original coursing lines of ashlar: an unacceptable quilted appearance results.

Now, however, through the transfer of technology from the marine engineering field, English Heritage and its consultants and contractors have been able to develop a system of remote surveying, key-hole surgery and preventive maintenance to limit the need for such damaging interventions by the employment of *cathodic protection*.

In its first application to an historic structure in the United Kingdom, at Inigo Jones's Gateway (in the ornamental grounds of English Heritage's Chiswick House in west London), an entire stone facade has been protected and its welfare continues to be electronically monitored.

This pioneering work included the development of a new class of super-sensitive metal detectors and the improvement of non-destructive survey techniques to accurately map and locate buried metal cramps in buildings.

THE CONSERVATION OF ARCHITECTURAL METALWORK

The conservation of the lead sphinx statue at Chiswick House and development and long-term testing of methods to clean and coat architectural wrought ironwork located in a marine environment are two case studies are included here to demonstrate the very high standards of conservation practice English Heritage now expects in the field of architectural metalwork.

At Chiswick House, a precious lead Sphinx has been sensitively consolidated and re-presented indoors for its protection. At Portsmouth Garrison Church, another English Heritage property, what appear at first sight to be common wrought- and cast-iron boundary railings have been properly recorded, assessed and conserved. So what is unusual?

It is a curious anomaly of international conservation practice that historic *stone* monuments receive far more attention than do their *metal* cousins. Iron, bronze and lead work have all been seen as transitory materials, because of the historic problems of maintaining and repairing them. Consequently, while trends in the conservation of other inorganic materials have moved technically towards the maximum retention of the authentic, original material, the metals field is still for the most part inhabited by those who would rather chop out decaying metalwork and replace it anew.

Pioneering developmental work at English Heritage, first by Nicola Ashurst and more recently in concentrated form by Bill Martin and his team in the Architectural Metalwork Conservation Studio in Regents Park, London have broken this pattern and set new standards of care that can now be generally applied. The final work of the metals conservation team (which closed in April 1997), of Andy Smith, Keith Blackney, Alun Walker, Nancy Hudson-Rooney and Jim Kennedy, is published here as a model for future practice.

Part I

Research

The behaviour of structural cast iron in fire
A review of previous studies and new guidance on achieving a balance between improvements in fire protection and the conservation of historic structures

ALAN PORTER
Warrington Fire Research Consultants, Airport House, Purley Way, Croydon, London CR0 0XZ
CHRISTOPHER WOOD*, JOHN FIDLER AND IAIN MCCAIG
English Heritage, 23 Savile Row, London W1X 1AB

Abstract

The behaviour of structural cast iron in buildings in fires has been little understood and subject to ill-informed prejudices over the years. Consequently passive fire protection methods employed when rehabilitating historic structures have been onerous and often detrimental to the special architectural interest and appearance of listed buildings. This paper reviews previous studies and fire tests which have enabled a more sophisticated fire engineering approach to be taken towards risk assessment. Guidance is offered on consequent fire protection methods that can strike a benign balance in building conservation terms.

Key words

Cast iron, fire, performance, tests, risk assessment, passive protection

INTRODUCTION

Cast iron was an important structural component of many buildings constructed in the late eighteenth and nineteenth centuries, particularly those erected for industrial and storage uses. Typical examples include mills, warehouses and factories, the best of which are now statutorily listed for their architectural or historical interest. Many cast-iron structures are now redundant but others have been, or are being, converted to new uses.

The process of rehabilitation is endangering the physical integrity of many historic structural elements on rudimentary safety grounds. Under current building regulations,[1] if a significant change of use is proposed, such as the conversion of the property to multi-occupied residential accommodation, or to office or public assembly functions, or if a substantial part of the envelope has to be reconstructed, the building would almost certainly have to be brought up to the latest construction standards, particularly in regard to passive fire safety. This can lead to the addition of fire protection to the cast iron resulting in the loss of significant architectural detail, severely damaging the unique character of the historic building. This is ironic, since the graceful lines of the construction were very often originally designed to be fire-proof.

Some of the detailed requirements imposed under the various safety codes relate not so much to previously established, scientifically derived performance results for individual materials, but rather to accepted norms based on perceptions and not facts. In particular, cast iron's performance in fires has often been called into question, in terms of early localized structural failures and thermal shock following dowsing during fire fighting, that have led to onerous measures being imposed that may have been unwarranted in many cases.

This paper examines why cast iron gained a poor reputation among safety regulators. It provides objective information on the actual, rather than perceived performance of cast iron in fire and concludes that concerns have often been unjustified. The paper also gives guidance on how fire protection can be provided, where essential, with the minimum loss of architectural quality and integrity.

Some of this information has been circulating among specialists for a number of years in published form (Barnfield and Porter 1984, 373-80), but it has never been supplemented and drawn together in the context of establishing a fire safety risk assessment strategy to ameliorate the architectural impact of potential changes to historic cast-iron construction. The paper will be of interest not only to structural engineers, fire engineers and fire brigade fire prevention officers but also to architects and building surveyors, local planning authority conservation officers and building control staff.

HISTORICAL PERSPECTIVE

Cast iron was first produced in the early sixteenth century following the advent of the blast furnace, although output was limited to small castings. Improvements in smelting by Abraham Darby in the early eighteenth century, using coke instead of charcoal, enabled castings of sufficient size and quality for structural purposes to be manufactured. This led to the increasing use of cast iron in large structures. Notable early examples include Abraham Darby's Iron Bridge over the river Severn at Coalbrookdale (1777-81), William Strutt's 'fire-proof' cotton mill in Derby (1792) and Charles Bage's five-storey Flax Mill at Shrewsbury (1796-7) (see Johnson and Skempton 1956, Trinder 1973 and Sutherland 1976). From 1800 onwards, cast iron was used extensively but not exclusively for pillars and beams in mills and warehouses. Cast iron was used in composite construction with timber and with wrought iron but, by the middle of the century, lost ground to the latter material, which had greater tensional strength.

It finally gave way to mild steel, which has greater ductility and higher compressive and tensile strengths, in the 1890s. Virtually no new structures have been built in cast iron this century, making it of marginal commercial interest,

* Author to whom correspondence should be addressed

which helps to explain why so little research and testing has been carried out, particularly on its performance in fires.

PROPERTIES OF CAST IRON

Cast iron has a high compressive strength but is low in tension. It was used predominantly for columns and to a lesser extent for beams: the bottom flange of the latter being relatively thick in order to provide extra tensional strength. The characteristics of cast iron are attributable to its impurities, particularly carbon and less so to phosphorous and silicon, introduced during smelting. These impurities helped the casting process but made the iron brittle, making it tend to break without warning when it failed, rather than gradually yielding through plastic creep as steel does.

The quality of manufacture is just as important as the inherent material properties in determining the strength of a cast-iron member. Casting processes were relatively erratic and sometimes produced air blow-holes in the iron or significant differences in wall thicknesses of hollow members. With these variations in the quality of manufacture the strength of a member could vary widely. Therefore large factors of safety have always had to be applied by engineers when determining the strength of the material (see GLC 1976).

THE BEHAVIOUR OF CAST IRON IN FIRE

Structures fail in fire for a number of reasons. Those pertinent to this paper are that:

- if the fire heats the structure to a high enough temperature for the metal to suffer plastic deformation, the strength of the members will be reduced to a point where they cannot support the applied load and will fail
- the metal expands significantly at high temperature in a fire and the expansion (eg of an iron beam) could cause the masonry wall upon which it is supported to be pushed out and collapse. The expansion of parts of a building may also set up large internal forces in the structure, possibly precipitating catastrophic failure of other members.

Cast iron loses its strength at high temperature marginally more slowly than steel, but they both expand by about the same amount when heated. At about 600° C, cast iron loses half its strength, whereas steel is affected at a temperature of approximately 550° C. A ten metre length of either metal will expand by about sixty millimetres when heated to these temperatures. In a fire, the stresses caused by this expansion are critical for cast iron, as any distortion can lead to the development of tensile stresses.

There is a general perception that cast iron has a poor reputation for its performance in fire. The main reasons cited are:

- inherent weaknesses in the cast iron caused by an unreliable production process
- inadequate design and control during construction resulting in over-stressed or eccentrically stressed columns and beams
- the likelihood of cracks occurring in a fire, owing to its brittle nature, causing a sudden collapse of the structure. This is believed to be often precipitated by the thermal shock caused by fire hoses being directed onto a hot column or beam during fire fighting.

Doubts, confusion and contradictions about the performance of cast iron as a non-combustible and 'fire-proof' product have been raised by a number of authors since the late nineteenth century. For example in the early 1880s *The Fireman* magazine was particularly critical of the idea of storing highly combustible materials in warehouses where the cast iron was unprotected. Yet in 1903 Freitag stated that, as a result of tests and actual fires, unprotected cast iron could stand temperatures of 1500° F virtually unharmed, while also carrying very heavy loads, despite frequent applications of cold water while the metal was at red heat.

By 1913, H G Holt in his book *Fire Prevention in Buildings* (1913) stressed that cast iron should receive protection against fire, it being 'incapable of permanent extension, it fails suddenly by tension and transverse stress when its limit of elasticity is reached. Cast-iron columns have failed badly in numerous fires, partly owing to the fact that the usual method of casting them, ie horizontally, may often leave small unnoticed defects which become dangerous points of weakness in a fire, particularly if water from a hose should catch them when they are heated'.

Following the major fire in San Francisco in 1921, Freitag contradicted his earlier conclusions and observed that the collapse of structures resulted from inherent weaknesses in the unprotected cast-iron columns, originally caused by the cooling and shrinking of the material when cast (Freitag 1921). Many instances were found where he observed such weaknesses, under the additional stress of heating and cooling, apparently causing the heads of cast columns to break off.

THE ARCHITECTURAL IMPACT OF FIRE SAFETY MEASURES ON CAST-IRON STRUCTURES

Faced with a paucity of technical guidance on the actual behaviour of cast-iron structures in fire, local building control and fire safety officers have had little alternative but to be cautious and set a prejudiced eye against the material, requiring a range of passive fire resistance measures to be taken to prevent catastrophic collapse. These interventions have differing degrees of impact on the architectural or historic interest of the structures involved: some are more acceptable than others from the conservation point of view.

In drastic cases, cast iron has been removed altogether and replaced with reinforced concrete that constitutes a complete loss of historic fabric. Almost as severe, though sometimes handled well architecturally, is the setting of a new fire-proof structure alongside the cast iron, as a failsafe in case of the latter's collapse during a fire. Most treatments, however, involve over-cladding with fire-proof material so that the structure is concealed or obscured and the original slender design fattened in the process, thus unacceptably affecting the character and appearance of the interior spaces in the building as well as the historic structural components that help to frame them.

The coating of cast iron with intumescent paints or papers may be less drastic, but this sometimes obscures architectural detail. Mouldings, profiles and sections can thus be obliterated or made ill-proportioned and the smooth texture of the painted ironwork is lost to the thick interwoven fibrous overskin. Another, less obtrusive, treatment involves grouting and filling the hollow void inside cast-iron columns with concrete. A more interventionist development of this last theme involves partially dismantling the historic structure to insert reinforced concrete inside the cast-iron columns.

Without improved guidance to assess cast-iron structures for their fire performance characteristics, it is extremely difficult for specifiers to choose the most appropriate and benign level of intervention from a conservation standpoint. Guidance to local planning and building control authorities in Planning Policy Guidance Note 15 (Department of the Environment and Department of National Heritage 1994) encourages a flexible and sympathetic approach to be taken to balance public safety and conservation interests. But in relation to concerns about historic cast-iron structures they also need specialist technical guidance on what may be necessary (or non-essential) from a fire safety point of view.

PERFORMANCE OF UNPROTECTED CAST IRON IN EXPERIMENTAL FIRE TESTS

Important tests have been carried out on the performance of unprotected cast iron in fire this century in the United States of America, in the United Kingdom and in Germany. In general, the results of the tests support the view that while the material can maintain its structural function in fire, it is prone to brittle failure.

In the USA a series of tests in the early part of the century were carried out on five randomly selected, unprotected cast-iron columns (Barnfield and Porter 1984). Three columns fractured, even though no water was applied at the end of the heating period. In the other two tests, water was applied to the heated specimens but no cracking was observed. None of the columns suffered significant deflection prior to the application of the hose stream.

The German tests (Barnfield & Porter 1984) were similar to those of the Greater London Council (see below). The tests were carried out on three randomly selected, full-sized columns and seven 'stumps' formed from material salvaged from demolition sites. The columns and stumps varied in size and in the quality of casting. All the stumps cracked during the tests but no reference was made to the full-sized columns cracking.

In the United Kingdom, five tests were carried out in 1982 and 1983 for the Greater London Council (GLC): three on columns and two on beams (Fig 1). The test method was in accordance with the standard test at the time (BS 476: BSI 1987) except that a hose stream was applied to the test specimens after the heating period. The columns and beams were loaded to the maximum stress permitted.

Two out of the three columns cracked after the application of the water. The other column did not crack even after the hose stream. Both of the beams cracked on the top (compression) flange ten minutes after the application of the hose stream had finished and after the load had been removed. All the columns and beams supported the reload applied 24 hours after the test despite being badly cracked. The results of these tests are summarized in Table I.

Figure 1. A cast-iron column being subjected to a water hose stream test after its release from the furnace at the Loss Prevention Council during the Greater London Council's experiments in 1982/3 (Photograph: Iain McCaig).

The report of the GLC tests by Barnfield and Porter (1984) came to these conclusions:

- Unprotected cast-iron elements can suffer from brittle failure under fire conditions but they can also achieve a standard of fire resistance in excess of 60 minutes and therefore should not be treated on the same basis as modern structural materials.
- In areas where it can be assured that the fire load and potential fire severity can be maintained at a low level, the use of unprotected cast iron or a reduced level of fire protection can be considered.
- Where a low fire load cannot be guaranteed, cast iron should be protected so that the temperature of columns or beams together with any connected iron or steel members should not exceed 300° C.
- If cast-iron columns are used in conjunction with timber beams, a failure temperature of 550° C can be assumed provided that the fire protection is resistant to rapid erosion by fire-fighting operations.

The question then arose, how to model the physical properties of cast-iron sections without the need to test each

Table 1. Experimental fire tests carried out in 1982-3

	Column or beam details	Applied load: tonnes (t), (stress N/mm²)	Hose application	Fire resistance time in mins & temperature °C	Comments
TE 3432 (FRS sponsored test)	Protected hollow column with 25 mm mineral wool (Density 60 kg/m³) Exposed length = 2170 mm OD = 127 mm ID = 76 mm	32 t (39)	yes	97 mins 510°C	No sign of cracking of column. Reload supported 24 hours after test
TE 4495 (GLC Column 3)	Unprotected hollow column Exposed length = 2170 mm OD = 127 mm ID = 76 mm	32 t (39)	yes	45 mins 750°C	Column cracked at mid height around the circumference 5 minutes after the end of the hosing down period. Reload supported 24 hours after test
TE 4496 (GLC Column 2)	Unprotected hollow column Exposed length = 3070 mm OD = 115 - 145 mm ID = 80 mm	37 t (44)	yes	30 mins 660°C	On hosing, two annular cracks extended around 25% of the circumference of the column. Reload supported 24 hours after test. The head fell off the column when it was removed from the furnace.
TE 4497 (GLC Column 1*)	Unprotected hollow column Exposed length = 2725 mm OD = 206 - 230 mm ID = 172 mm	107 t (74)	yes 22 mins	35 mins 630°C	Hosed for 22 minutes. Column bowed away from water jet. No cracking was observed. Reload supported 24 hours after test.
FIRTO 7-9-79 (FRS sponsored test)	Protected hollow column with 50 mm mineral wool Exposed length = 2133 mm OD = 127 mm ID = 76 mm	32 t (39)	yes	135 mins 750°C	No cracks were observed after application of the hose stream. Reload supported 24 hours after test.
FIRTO 26-11-79 (FRS sponsored test)	Unprotected hollow column Exposed length = 2133 mm OD = 127 mm ID = 76 mm	32 t 39	yes	42 mins 750°C	No cracks were observed after application of the hose stream. Reload supported 24 hours after the test. After 24 hours, the load was increased to 112 t and the column did not collapse.
FIRTO 3-8-81 (Alan Baxter and Associates)	Unprotected solid column from New Concordia Wharf, London Exposed length = 2640 mm Ornate square section of approximately = 115 x 115 mm	18 t (15)	yes	69 mins 810°C	No cracks were observed during the fire test or hose streaming. Reload supported 24 hours after the test.
WRCSI 31948	Unprotected beam Depth = 305 mm FW (top) = 90 mm FW (bottom) = 185 mm Wt/unit length = 116 kg/m Effective span = 4070 mm	23 N/mm² tension bottom flange	yes	63 mins	Hosing with water produced a crack in top flange and upper half of web. Reload supported 24 hours after the test. 861°C
WRCSI 30216	Unprotected beam Depth = 305 mm FW (top) = 90 mm FW (bottom) = 185 mm Wt/unit length = 116 kg/m Effective span = 4070 mm	23 N/mm² tension bottom flange	yes	65 mins 858°C	Hosing with water produced a crack in top flange and upper half of web. Reload supported 24 hours after the test.

OD = Outside Dimension   ID = Inside Dimension   FW = Flange Width

* The column dimensions given in this table are actual measurements taken after sectioning the columns. The load calculations were of necessity based upon *nominal* dimensions and, in the case of GLC column 1, this meant that the stresses generated were approximately 40% higher than specified in the 1909 London Building Acts.

This table is taken from J. R. Barnfield & A. M. Porter *Historic Buildings and Fire: fire performance of cast-iron structural elements* in the Structural Engineer. Vol 62A. No 12. London. December 1984. pages 373-380 and reproduced here with by kind permission of the Institute of Structural Engineers.

column in the field in order to assess its likely performance in fire situations? Here a short digression is needed.

In the late 1970s and early 1980s, research on the fire protection of structural steel work in the United Kingdom and Germany moved towards devising a universal answer: fire damage susceptibility factors became employed using linear regression analysis to calculate the risks.[2]

In the model, a structural section with a large perimeter will receive more heat in a fire than one with a smaller perimeter. Also, the greater the cross-sectional area of the section, the greater the thermal capacitance of the member involved. It therefore follows that a small thick section will be slower to increase in temperature than a large thin one. A factor incorporating measurements in this way gives an indication of the member's rate of temperature gain during a fire: the quicker it heats up, the sooner it reaches its plastic condition and starts to deform or structurally fail. So the higher the value of the factor, the greater will be the thickness of fire protection necessary for the member to resist fire for a given time (Association for Specialist Fire Protection 1988, 6-9).

To take account of the varying shapes and integrity of cast-iron structural members, this so-called section factor, H_p/A, can be employed. The ratio consists of an estimate of the H_p or heated perimeter area (ie circumference x length), over A (the cross-sectional area) and, as stated above, provides a measure of the inherent fire resistance of a beam or column. Put another way, the heated perimeter is a measure of the exposed area which determines the rate at which heat enters the member and the cross-sectional area is related to the mass which determines the amount of heat required to heat the member to its failure temperature. The lower the H_p/A ratio, the greater the inherent fire resistance of the cast-iron structure.

Using the model, the GLC tests concluded that unprotected cast-iron columns with an H_p/A factor of less than 45m^{-1} can be expected to achieve a 30 minute standard of fire resistance and this would therefore meet fire resistance standards for a wide range of applications.

PERFORMANCE OF CAST IRON IN REAL FIRE SITUATIONS

Following its earlier tests, in 1986 the GLC commissioned the structural engineers Alan Baxter and Associates to investigate the performance of cast iron in real fires. This additional work was initiated because it was felt that there was anecdotal evidence that the probability of failure of cast iron in fires was much lower than that indicated by the experimental fire tests, where most of the columns and all of the beams suffered structurally significant cracking. A number of fire brigades in the UK and the New York City Fire Department were contacted and records of fires going back to the 1800s were searched out and reviewed.

The general opinion of the fire officers was that cast iron performs well in fires and is much less prone to collapse in the early stages of a fire than unprotected steelwork. However, it is always treated with respect by firemen who, by experience and training, are wary of the buildings in which it is found. Although it was generally expected that a cast-iron structure would collapse if a fire caught hold, this was

Figure 2. A slender cast-iron column remains standing after the fire which destroyed the Empire Theatre, Stoke-on-Trent in January 1993 (Photograph: English Heritage).

thought to be due to the general lack of fire protection measures and not always to do with the cast iron itself. Structural collapse, for example, was often initiated by the movement of machinery as a result of the burning through of a timber floor or failure of a beam.

Multi-storey buildings with brick jack arch floors, cast-iron beams and columns have been known to survive major conflagrations. They remain standing despite the loss of columns or the cracking of beams and are often repaired and refurbished thereafter. The report (Baxter and Associates 1986) had these conclusions:

- (Fig 2) Cast-iron columns can occasionally crack but they generally perform well in fires. Where failure does occur, it is usually limited to one or two columns and not to the entire set across the building. Often cast-iron columns are the last structural members left standing after a fire (Figs 3 and 4).
- Where cast-iron members have failed, the reasons have included the variability of the material, quality of casting, working stress, slenderness and a high H_p/A ratio.
- Cast-iron beams are more prone to sudden failure and collapse than columns, especially if a fire-fighting water hose stream is applied to them when already very hot.
- In general there were no fundamental objections to the principle of leaving cast-iron columns unprotected in areas of low fire risk. However, owing to the variability of the

Figure 3. A cast-iron column shows little distortion after a fire. The Empire Theatre, Stoke-on-Trent (Photograph: English Heritage).

size, shape and quality of cast-iron members and of their constructional configuration, it is impossible to give any universal rules where cast iron could be left unprotected.
- In each case, before a decision could be made on a building, consideration would have to be given to factors such as occupancy, means of escape, fire load, compartmentation, early warning of fire, other means of fire protection, access for fire fighting, quality of construction and the detailing and the overall robustness of the building.

Although not expressly referred to in this report, these factors are all now essential given components of fire safety engineering exercises applied in complex and sensitive cases.

SUMMARY

There is clearly some conflict between the traditional perception of the behaviour of cast iron in fires, which is supported by the results of the experimental fire tests, and the more recent assessment of field experience of real fire situations. This could be due to a number of factors.

The historic view was probably influenced by the poor quality of many of the buildings where the cast-iron members were very highly stressed and/or of poor quality. Alternative contemporary views may be influenced by fires which have occurred in unused buildings and where there was very little applied load on the cast-iron members. Modern fire-fighting techniques are also far more sophisticated and fires can now be dealt with far more quickly and efficiently, resulting in whole structures being saved.

However, the effects of refurbishing and upgrading buildings to comply with current safety regulations could, ironically, make the fire behaviour of cast iron unpredictable. For example, upgrading floors to achieve a specified period of fire resistance could result in a more severe fire by limiting the ventilation and reducing the heat loss. This could adversely affect the cast iron.

The general conclusion is that unprotected cast iron should be treated cautiously, but where the fire load is low, and the cast iron can be shown to be of good quality, it should not be necessary to provide additional fire protection. Fire engineering methods should be used to model and evaluate the effects of a fire on cast-iron members in individual buildings. By such means, the potential severity of a fire in a building can be estimated and the temperature likely to be reached by the cast-iron structural members can be calculated. Guidance on how fire engineering methods can be used is to be found in BS 5950 (BSI 1990).

The current UK Home Office advice to fire brigades has taken all the research efforts of the 1980s into account and offers now a considered, balanced line. It states that 'besides the very real danger of distortion and collapse of all metals under heat, cast iron has always had a bad reputation for its behaviour in a fire. It is open to doubt whether this reputation is fully justified for, after many serious fires, cast-iron columns have been found in place when the steelwork has collapsed' (Home Office 1990).

CONCLUSIONS

The cast-iron structures of historic buildings are valuable cultural resources that reflect the technical ingenuity and craftsmanship of a bygone age. Many structures are now officially recognised for their importance by being statutorily listed as of national architectural or historic interest and works that materially affect them are subject to listed building consent procedures under planning control (Planning [Listed Buildings and Conservation Areas] Act 1990).

But there is a strong perception in the fire-fighting, building control, engineering and specifying communities that cast-iron columns and beams are highly susceptible to structural cracking and catastrophic failure in fire situations, especially where fire hose water is played onto heated metal during fire fighting. There are even anecdotal stories told of cast-iron columns exploding in fires.

These perceptions have been reinforced by a limited interpretation of experimental fire test data on a statistically limited number of randomly-selected samples. However, a detailed evaluation of the historic incidences of cast-iron failures in fire, of fire tests here and abroad and of more recent cases of fires in cast-iron buildings witnessed in the field (as reported by fire officers themselves), as set out above, has concluded that the concerns are genuine but often exaggerated. Consequently:

Figure 4. Only one of the cast-iron beams shows any signs of cracks after the fire that destroyed this building in Kensington Palace Gardens in 1991 (Photograph: English Heritage).

- Cast iron can be left unprotected where the fire loads are proven to be limited and the cast iron is found to be of good quality, or where the material is used in composite construction, ie of cast-iron columns and timber beams, where the structural shear forces during fire situations are usually minimal.
- A fire safety engineering solution should be considered to determine the degree of, and to devise a reduction in, the fire loads and risks concerned. This approach will help define the necessity for and minimum fire resistance of the fabric. The exercise should be carried out by a qualified fire engineer experienced in carrying out this type of analysis in historic buildings.
- Where there is a proven risk to the structure, active and passive fire protection measures may be combined to ameliorate the need to carry out excessive works to protect the cast iron from fire and/or dowsing which may jeopardise or compromise the special architectural or historic interest of the structure. For example, improvements in building management that eliminate or reduce fire hazards, or the installation of early fire detection and alarm systems tied to local trained firefighting or localized automatic fire suppression systems, may eradicate some of the needs for improvement of the passive fire resistance of the structure.
- Where specific changes for increased passive fire resistance are necessary, the available treatments can be ranked according to their impact on the special architectural or historic interest of the cast-iron structure and its setting and are recommended as follows in descending order of merit: filling with concrete, coating with intumescent paint, concrete filling and intumescent coatings combined, cladding in mineral wool and providing an adjacent, alternative, failsafe structure.

Techniques for retaining or improving the fire resistance of cast-iron structures

Assessment: checking the quality and integrity of the cast iron.
Before deciding upon the necessity for and then types and degree of active and passive fire protection for an historic cast-iron structure, it is important to measure all the relevant parameters which might affect its performance in a fire situation. For example, the structural integrity of the cast iron should be evaluated *in situ*. It is a specialized procedure, not to be undertaken lightly (see Bussell 1997). The quality of the assessment will be influenced by the surveyor's knowledge of the material, details of its manufacture and possible inherent weaknesses, the effectiveness of the sampling technique and by the types of tools used in the survey.

Assessment regimes include locating voids and blow holes in the metal cross-section by vibration integrity testing, ultra-sound and radiography; and measuring anomalies in the thickness of walls in the material through the employment of ultra-sound, radiography and core sampling or a combination of techniques (see Fidler 1980 and Hollis and Gibson 1991).

Making structural repairs
There is a wide range of remedial treatments currently available for improving the overall structural integrity of cast-iron members *in situ*. Welding, metal stitching and strapping cast iron are all commercially viable, practical solutions, the details of which are beyond the scope of this report (Wallis 1988, 13-41).

Methods for upgrading fire resistance
FILLING WITH CONCRETE
Hollow column castings were sometimes designed to carry rainwater goods or to act themselves as rainwater courses. Their cap details also vary, depending upon the connections to beams overhead and often prevent or limit access to their interior for direct vertical runs, eg for the introduction of reinforcement. There will be only a limited number of occasions when it will be practical for columns to be filled with traditionally reinforced concrete.

However, by coring through the side, it is feasible to fill cast-iron columns with a simple cement grout. This method of upgrading, without reinforcement, is unlikely to achieve more than a 60 minute standard of fire resistance. There is no fire engineering guidance specifically available but there is extensive information on the fire performance of concrete-filled structural mild steel hollow sections such as that in BS 5950 (BSI 1990). The guidance should be applicable to the filling of cast-iron columns.

Some of the specimens in the German tests, referred to above, were filled with concrete in order to establish whether or not this improved their fire resistance. The concrete-filled columns performed better than the unprotected ones because the concrete core absorbed some of the heat from the cast iron and took up part of the applied load in compression. The report of the tests concluded that cast-iron columns of any H_p/A factor filled with high-grade concrete can be regarded as having a 30 minute standard of fire resistance and if the H_p/A factor is less than 23m^{-1}, it is possible to achieve a fire resistance of 60 minutes.

Concrete filling would enable the external appearance of the cast-iron member to remain relatively untouched with mouldings, profiles and surviving historic paint layers left *in situ*. Cores can be replaced and welded or metal stitched back in place after the upgrading has taken place. The technique thus provides a mostly benign and inconspicuous method of upgrading. The high level of subsequent alkalinity on the internal face of the column shaft would also be beneficial in preventing or limiting corrosion in damp conditions. However, the process is practically irreversible and there are a number of technical problems with its effective introduction.

For example, injecting the concrete can be difficult and experience with filling steel columns has shown that ensuring adequate compaction of the material can be a problem. Concrete tends to shrink with age thus reducing the effective cross-sectional area of the composite column. However, this process is likely to be very slow as the cast iron will provide an effective early barrier to carbonation of the concrete filling.

Drill holes may need to be made at intervals, initially to relieve the air pressure and allow the grout to flow. Hydrostatic pressure and the build-up of steam from the exothermic reaction within the concrete as it sets would also need to be relieved via vent holes in the column and subsequently made good. There is also no guarantee of escape for the excess water vapour associated with grouting which could lead to the triggering of localized corrosion. The long-term effects and impact of these sort of problems in the remedial treatment of cast iron have not been investigated.

COATING WITH INTUMESCENT PAINT

A more easily applied, potentially reversible and less costly form of upgrading is the external application of an intumescent coating to the exposed surfaces of cast iron. It is a treatment that has been adopted frequently, though not without technical and aesthetic problems (Bidwell 1980, 24–30). The subject is discussed in *The use of intumescent products to provide fire protection in historic buildings* (English Heritage 1997; see also McCaig and Porter 1988) from the conservation viewpoint, and in BS 8202 (BSI 1992) concerning technical issues for coating systems for metallic substrates to provide fire resistance.

There are a wide range of intumescent coatings available for the protection of cast iron. Some of these are relatively thin film coatings which can provide up to 90 minutes fire resistance as defined in BS 476 Part 20 (BSI 1987) with coating thicknesses up to 3 mm, dependent on the period of fire resistance required. For periods of fire resistance of more than 90 minutes, intumescent coatings can still be used but

Figure 5. Intumescent paint coating 6 mm thick applied to the cast-iron columns at the Covent Garden Market Building, London. Purpose-made profiled hardwood rollers were specified by the Historic Buildings Division of the Greater London Council to finish and shape the newly applied coating in an effort to improve the appearance of the newly covered mouldings and profiles (Photograph: Iain McCaig).

the thickness of coating required may be substantial, typically in excess of 6 mm to 10 mm, which can be extremely visually intrusive. In such cases it may be possible to balance the need for excessive thicknesses of intumescent coatings, that materially affect the special interest of the historic structure, against alternative or supplementary fire safety measures (Figs 5 and 6).

A disadvantage of intumescent coatings is that the iron will need to be stripped of all existing paint layers to ensure the effectiveness of the fire protection. There would thus be a consequent loss of historic paint and archaeological evidence, in which case consideration should be given to making a reference set of historic paint samples in advance of the work being implemented. Intumescent coatings have a relatively limited colour range and so cannot always faithfully replicate historic colours. They also cannot generally be overcoated with traditional decorative paints by way of camouflage, as their stability and/or performance in fire can be affected.

Guidance on the intumescent coatings commercially available can be found in *Fire protection for structural steel in buildings* (Association for Specialist Fire Protection [ASFP] 1988). Columns can be treated as being comparable to hollow steel sections with the same H_p/A factor when

Figure 6. The application of an intumescent coating has obscured the sharp arrises and detail at the base of this column at the Market Building, Covent Garden, London (Photograph: Iain McCaig).

selecting the thickness of protection necessary to provide the requisite standard of fire resistance. However, the intumescent coating should have been tested on a steel section which has a comparable shape and size to that of the cast-iron member to be protected, as it is possible that with very low H_p/A values the mass of the section may adversely affect the performance of the coating.

CONCRETE FILLING AND INTUMESCENT COATINGS COMBINED

Intumescent coatings can be used in conjunction with concrete filling of cast-iron columns either to provide a higher level of fire resistance than each method would provide individually, or a similar level of resistance without the necessity for very thick coatings that obscure historic details. However, fire tests may be needed before any definite conclusions on levels of protection can be reached. There is a possibility, for instance, that the absorption of heat by the concrete core may inhibit the intumescent coating from forming an adequate insulating layer as it foams. The fire resistance provided may therefore be less than expected.

CLADDING IN MINERAL WOOL

In general, this method of upgrading historic cast-iron construction would obscure the columns or beams to which it was applied and would therefore be unacceptable on conservation grounds. However, it may be appropriate for a limited, defined temporary period, for example in a mothballing exercise for a newly redundant building or during construction work as a means of dual protection from physical damage and fire, after which the cladding could be removed and more benign, permanent treatments applied where necessary.

The GLC report (Barnfield and Porter 1984) also contains details of fire tests with mineral wool cladding on cast-iron members. A column protected by a 25 mm thickness of mineral wool achieved a 97 minute standard of fire resistance and a column protected by a 50 mm thickness of mineral wool achieved a 135 minute standard of fire resistance.

PROVIDING AN ADJACENT, ALTERNATIVE, FAILSAFE STRUCTURE

There are likely to be very few circumstances where fire risks to cast iron cannot be ameliorated by a combination of methods described above, using both active and passive methods in a comprehensive fire engineering approach to historic buildings, that do not compromise the special interest of the structure. But there have been special situations where a radical alternative strategy has been adopted which leaves the historic cast iron untouched and does not rely upon its performance in fire in case of catastrophe.

This system employs a new alternative structure for the building using reinforced concrete, sheathed steel or masonry piers that comply with modern safety code requirements in terms of fire resistance and are designed to take up the loads in case of the collapse of the neighbouring cast iron. Carefully handled detailing and location of such structures give a subservient visual appearance and yet provide a structural failsafe for the historic material. This approach is obviously expensive and so has only been used in situations where other safety factors, such as floor loadings, have required large-scale improvement. This technique may be found in some warehouse conversions, for example in Liverpool.

ENDNOTES

1   See DOE and Welsh Office 1991 and Fire Precautions Act 1972.
2   In the UK, the work was undertaken by the Fire Research Station, in collaboration with members of the Association of Structural Fire Protection Contractors and Manufacturers (ASFPCM), the Greater London Council's Scientific Branch (later known as London Scientific Services) and the laboratories of the Loss Prevention Council and Warrington Fire Research Establishment. See Association for Specialist Fire Protection (ASFP) 1988. This is now an industry standard and is referred to in Appendix A of Approved Document B of the Building Regulations 1991.

BIBLIOGRAPHY

N. K. Allen, *The care and repair of cast and wrought iron*, unpublished dissertation for the Diploma in Conservation Studies, (York: York University, 1979).

J. Ashurst and N. Ashurst, *Metals*, Practical Building Conservation, English Heritage Technical Handbook Series, 4, (Aldershot: Gower Technical Press, 1988).

Association for Specialist Fire Protection, *Fire Protection for Structural Steel in Buildings* 2nd edition, (London: Association for Specialist Fire Protection [ASFP, formerly Fire Association of Structural Fire Protection Contractors and Manufacturers Ltd, ASFPCM], 1988).

J.R. Barnfield and A. M. Porter, 'Historic buildings and fire: fire performance of cast-iron structural elements', in *The Structural Engineer*, 62A, no 12 (December 1984), 373-80.

Alan Baxter and Associates, *The Performance of Cast Iron Structural Elements in Actual Building Fires*, (unpublished: Greater London Council, 1986).

T. G. Bidwell, 'The Restoration and Protection of Structural and Decorative Cast Iron at Covent Garden Market', in *Transactions of the Association for Studies in the Conservation of Historic Buildings* (ASCHB), 5 (1980), 24-30.

British Cast Iron Research Association, *Cast Iron in Building Structures: revived interest in a proven case*, Report No. X181, (Birmingham: British Cast Iron Research Association, 1984).

British Standards Institution, BS 476 *Fire Tests on Building Materials and Structures* Part 20: *Method for the determination of the fire resistance of elements of construction (general principles)*, (London: British Standards Institution, 1987).

British Standards Institution, BSCP 5950 *Structural Use of Steelwork in Building*. Part 8: *Code of practice for fire resistant design*, (London: British Standards Institution, 1990).

British Standards Institution, BSCP 8202 *Coatings for Fire Protection of Building Elements*.Part 2: *Code of practice for the use of intumescent coating systems to protect metallic substrates for providing fire resistance*, (London: British Standards Institution, 1992).

M. Bussell, *Appraisal of Existing Iron and Steel Structures*, SCI Pub 138, (Ascot: The Steel Construction Institute, 1997).

Department of the Environment and Department of National Heritage, *Planning and the Historic Environment*, Planning Policy Guidance Note 15, (London: HMSO, 1994).

Department of the Environment and The Welsh Office, *The Building Regulations 1991 Part B* (London: HMSO, 1991).

J. A. Fidler,'Non-destructive Surveying Techniques for the Analysis of Historic Buildings', in *Transactions of the Association for Studies in the Conservation of Historic Buildings* (ASCHB), 5, (Kilmersdon: ASCHB, 1980), 3-10.

English Heritage, *The Use of Intumescent Products in Historic Buildings*, (London: English Heritage, 1997)

J. K. Freitag, *Fireproofing of Steel Buildings*, (New York: John Wiley & Sons, 1903).

J. K. Freitag, *Fire Prevention and Fire Protection* 2nd edition, (New York: John Wiley & Sons, 1921).

Greater London Council, Department of Architecture and Civic Design, 'Cast-iron columns and beams', in *Development and Materials Bulletin*, 91, (London: GLC, 1976), 7/1-7/A2.

M. Hollis and C. Gibson, *Surveying Buildings* 3rd edition, (London: Royal Institution of Chartered Surveyors, 1991).

H. G. Holt, *Fire Protection In Buildings*, (London: Crosby Lockwood and Son, 1913).

Home Office, *Manual of Firemanship. Book 8: Building Construction and Structural Fire Protection*, (London: HMSO, 1990).

H. R. Johnson and A. W. Skempton, 'William Strutt's Cotton Mills 1793-1812', in *Transactions of the Newcomen Society*, XXX (1956).

I. McCaig and A. Porter, 'The use of intumescent coatings to provide fire protection in historic buildings', in *Transactions of the Association for Studies in the Conservation of Historic Buildings* (ASCHB), 13, (Kilmersdon: ASCHB, 1988), 47-50.

R. J. M. Sutherland, 'Pioneer British Contributions to Structural Iron and Concrete', in C. E. Peterson, *Building Early America*, (Radnor, Pa: Chilton Book Co, 1976), 96-118.

B. Trinder, *The Iron Bridge*, (Telford: Ironbridge Gorge Museum Trust, 1973).

G. Wallis, 'The Repair and Maintenance of Cast Iron and Wrought Iron', in J. Ashurst and N. Ashurst, *Metals*, Practical Building Conservation, English Heritage Technical Handbook Series, 4, (Aldershot: Gower Technical Press, 1988), 13–41

ADDRESSES

Association for Specialist Fire Protection, 235 Ash Road, Aldershot, Hampshire GU12 4DD;tel: 01252.21322.

English Heritage free leaflets, available from English Heritage Customer Services, 429 Oxford Street, London W1R 2HD; tel: 0171 973 3434.

Steel Construction Institute, Silwood Park, Ascot, Berkshire SL5 7QN; tel: 01344 23345.

ACKNOWLEDGEMENTS

Table 1 is reproduced with the kind permission of the UK Institute of Structural Engineers.

AUTHOR BIOGRAPHIES

Director of Warrington Fire Research Consultants (London), **Alan Porter** (MSc, MIFireE, MIFS) has been involved with fire safety in buildings for over twenty years. Prior to joining his present company he worked for the Greater London Council Scientific Branch. He has specialized in advising on fire safety in a wide range of complex buildings, including historic buildings, where standard guidance cannot readily be applied. In recent years he has advised on fire safety at a number of major historic sites, including royal palaces. Clients include English Heritage and the Historic Royal Palaces Agency.

Christopher Wood (BSc, AA Grad Dipl Conservation, MRTPI, ARICS) is a chartered surveyor and town planner with a specialist postgraduate qualification in building conservation. He is a senior architectural conservator with English Heritage, responsible for providing technical advice and for the management of several research projects, including those on fire safety in historic buildings (until 1996). He was also responsible for the daily running of the English Heritage Building Conservation Training Centre at Fort Brockhurst, Hampshire. Previous experience was gained in private sector architectural practice specializing in building conservation and as a conservation officer in local authority planning departments.

John Fidler (DipArch, MArch, MA Conservation, AA Grad Dipl Conservation, RIBA, FRSA) is Head of Architectural Conservation at English Heritage and is responsible with others for developing technical policy, providing technical advice, for research and development work on building materials decay and their treatment, for technical training and for the organisation's outreach campaigns. He was the first Historic Buildings Architect for the City of London Corporation, the first national Conservation Officer for Buildings At Risk, and the youngest and last Superintending Architect in the public service for the conservation of the country's historic estate. He has long been interested in fire safety in historic buildings and has contributed to English Heritage's policies on disaster preparedness and fire safety over many years.

Now working for John Renshaw and Partners Architects in Edinburgh, **Iain McCaig** (DipArch) spent several years in the Research and Technical Advisory Service of English Heritage and later in its Architectural Conservation team where he specialized in fire safety measures for historic buildings. He was instrumental in setting up disaster preparedness strategies for the organisation's historic house museums and published several papers on fire safety, leading English Heritage's response to the Bailey Report in the aftermath of the Windsor Castle fire. His previous experience was formed with the Architects Department at the London Borough of Camden, with the Historic Buildings Division of the Greater London Council, with the Conservation Practice on National Trust sites and as a consultant to archaeological excavations in the West Midlands.

The underside corrosion of lead roofs and its prevention

BILL BORDASS
William Bordass Associates, 10 Princess Road, London NW1 8JJ

Abstract

This report describes the research project on underside lead corrosion and sets out the findings to date. It covers the earlier work of the 1980s and early 1990s but concentrates mainly on the research programme that was commissioned by English Heritage from 1993–6. This work was needed because of the suspected increase in the occurrence of underside corrosion on historic buildings, with the failure of conventional theories to explain what was happening. The aim of the research was to understand the mechanisms that give rise to corrosion and consider the implications for existing materials and working practices in order to try and prevent its recurrence and minimize the amount of alteration needed to building fabric. Testing was carried out in laboratories, on purpose-made rigs and on a wide range of sites and these are described along with initial conclusions. Much of the information in this report formed the basis of the joint English Heritage/Lead Sheet Association's advice note for specifiers.

Key words

Lead sheet roofing, condensation, corrosion, roof design, treatments

EXECUTIVE SUMMARY

For centuries plumbers have known that lead roofs are susceptible to underside corrosion, but this appears to have been regarded as a normal part of the decay of a nevertheless durable material, with a lifespan to re-casting of typically fifty to one hundred years or more. In the late 1970s, however, awareness and concern about underside lead corrosion increased.

Work in the 1980s, particularly by the Ecclesiastical Architects and Surveyors Association (EASA) and the Lead Development Association (LDA) led to:

- identification of the main problem as the attack of lead by condensed moisture
- promotion of better ventilation underneath the lead
- recommendations against warm roof construction in which moisture could be trapped
- development of ventilated warm roofs for new buildings and major alterations.

By the early 1990s the situation had improved but was not entirely resolved. Some roofs subject to condensation were found to be performing well, while better-ventilated roofs (both ventilated warm roofs and traditional cold roofs with ventilated air spaces) were not always entirely free from underside corrosion. Continuous monitoring also indicated that most corrosion occurred not when the lead was wet but while it was drying out, and site tests indicated that conditions encountered by the lead early in its life might significantly affect its long-term behaviour.

The conclusions for historic buildings were unclear. Given the architectural and historic importance of lead and the principle of minimum intervention, reconstruction as a ventilated warm roof would always be a specification of last resort, but when was it really necessary, how should it be specified and how could its impact be minimized?

After preliminary studies over some five years, in 1993 English Heritage funded a programme of research and investigations by the corrosion engineers Rowan Technologies Ltd (RTL), involving theoretical work, laboratory and full-scale tests and site studies. This ongoing project is also supported by the Historic Royal Palaces Agency (HRPA) and the Lead Sheet Association (LSA). At the same time, English Heritage appointed William Bordass Associates to assist with the technical management of RTL's contract, to liaise with relevant English Heritage advisory cases and other research and to help bring together and present the results.

Summary of findings

At the start of the study, it was thought that the main cause of the perceived increase in underside lead corrosion was an increase in condensation dampness under the lead. This was attributed to changed environmental conditions in buildings and roof spaces through alterations to heating, ventilation, insulation, occupancy and control. However, the research has revealed a more complex set of mechanisms.

Moisture, particularly condensation, is a crucial agent in the corrosion of lead. Their relationship is not straightforward and some roofs known to be subject to condensation show little or no underside corrosion. The extent and temperature of wet/dry cycling are also important, as moisture evaporates and re-condenses, for example in intermittent sunshine. Organic acids are often also present and exacerbate the attack.

Conventional condensation analyses are of limited use in determining susceptibility to underside lead corrosion. Most traditional lead roofs fail these condensation checks and today's good practice design principles, but many have performed well in practice. The normal models also focus on diffusion of water vapour from inside the building in winter, while in practice the passage of moist air is usually more important. Dynamic transient conditions at the lead/substrate interface may also affect corrosion more than the seasonal build-up normally calculated: these include condensation from outside air on still, clear, dewy or frosty nights when the roof surface temperature falls below outside air temperature, and refluxing trapped or ingressed moisture with changes in temperature and solar radiation.

Ventilation of the roof space and the underside of the lead can help to avoid underside lead corrosion. Principles can be difficult to apply and are frequently misunderstood. There are two main mechanisms: 1, ventilation by outside air; 2, ventilation to the underside of the lead.

The dynamic performance of roof spaces with limited ventilation and containing large quantities of hygroscopic buffer material such as timber can provide some protection from corrosion.

Acids from the underlying timbers can greatly increase damage in damp condensing and refluxing environments. These acids are not consumed during the corrosion process, but act as catalysts which are continuously regenerated, and may indeed build up in concentration over time. Manufactured boards such as plywood, hardboard, chipboard and oriented strand board contain acids from constituent timber species, from glues, and may also have been hydrolysed during processing. When these materials get damp, underside lead corrosion can be particularly bad, so they are not recommended for most purposes. Timber preservatives can also cause corrosion, particularly if hygroscopic or if the wood has not been properly dried out before being used. While fresh concrete, mortar and lime are corrosive to lead, once aged and carbonated they can have a protective effect.

Dynamic modelling suggests that the hygrothermal properties of substrate timbers can significantly influence the amount of condensation. Taken together with their chemical properties, this makes a case for selecting timber deckings very carefully.

Under RTL's accelerated laboratory test conditions, no significant differences have yet been found in the susceptibility to underside lead corrosion of clean samples of sand-cast or milled lead, of modern and historic lead or of different chemical compositions. Continuously-cast (DM) lead is now being studied.

Research has found that initial surface conditions can have a significant influence on long-term behaviour. Fresh clean lead will start to corrode at its first encounter with moisture, and after that it is much more difficult for subsequent protection to form. However, lead can be protected from a succession of condensation/evaporation cycles by passive films built up either:

- on site
- exposed outdoors for two to three months and using the weathered topside as the underside
- in the laboratory, when lead is exposed to moist conditions close to the dewpoint but in which little condensation occurs
- in the laboratory by treating the lead with suitable chemicals.

In environments subject to periodic condensation, site and weather conditions at the time of laying can significantly affect initial surface film formation and hence long-term corrosion behaviour. Lead laid in the autumn, or on wet boarding, can corrode immediately.

Laying lead on wet substrates invites initial corrosion. Dry substrates will not only prevent this but suitable timbers (including some pine species), if laid dry or having dried out in the summer, may provide significant protection well into the winter.

Unfortunately spontaneously-formed films cannot be created reliably and the pre-formed ones vary in performance and will be damaged by working and handling on site. Simple ways of passivating or protecting the lead *in situ* have therefore been sought, using products which are widely-available and relatively safe. The most successful of these has been painting the underside with a slurry of chalk powder dispersed in water. This forms a passive film within ten minutes at room temperature. Extended site tests are continuing to determine how much long-term protection this may afford.

Accelerated testing indicates that such passive films can survive 50 or more condensation–evaporation cycles. Preliminary theoretical studies by the Building Research Establishment (BRE) suggest that lead in contact with a timber substrate will not normally encounter this many in the course of a year. However, passive films will break down eventually. Spontaneous repair may occur in moist conditions, particularly when hot, and calcium carbonate can promote this. In addition to the chalk slurry treatment, laboratory and site tests suggest that chalk left in place and/or in an impregnated underlay may provide continued protection. Development tests are continuing.

Underlays can have a significant effect on corrosion behaviour, but in damp situations the observed effects to date are mostly bad. Permeable fleeces improve access of air to the lead (good) but also let through moist air and water vapour (bad), assist drying-out (good), but during evaporation corrosion is faster (bad). Impermeable membranes potentially stop the ingress of moist air and water vapour (good): however, any ingressed water can be trapped (bad) and may then reflux (bad) in a low-carbon dioxide environment (often bad) in which acids may sometimes accumulate (very bad). Double-layer underlays (a lower layer to keep acids and condensation at bay and an upper one to look after the lead) looked promising but have been disappointing in tests. Underlays with controlled permeability and hygroscopicity are now being investigated, as are suitable underlays for the chalk coating process.

Provisional conclusions

For new roofs and major alterations the research endorses the concept of the ventilated warm roof but underlines the importance of attention to detail,[1] in particular:

- full ventilation from eaves to ridge with no dead spots
- a sealed air and vapour barrier below the insulation
- adequately-sized air spaces, ventilation inlets and outlets
- a substrate of low chemical reactivity.

Even where there is no additional water vapour, transient condensation in a ventilated warm roof can cause some underside lead corrosion, particularly over the gaps between the boards, and a second line of defence is desirable. While this has not yet been investigated exhaustively, in marginal cases plain building paper has been sufficient. Chalk treatments may give added long-term protection.

Roof space environments vary tremendously. While Dutch-barn-like environments with high ventilation rates of 100% outside air are one ideal, the research suggests that where air-seals and vapour-control layers separating the building from the roof space either do not exist or are of limited effectiveness, extra ventilation may sometimes be counter-productive. Unless they are dry, well-ventilated and preferably continuously heated, buildings without roof spaces are at high risk of condensation and underside corrosion.

To evaluate existing roofs the following procedures are suggested:

- If there is clearly severe condensation and associated dampness, timber decay etc, the situation needs careful review to define and correct these problems, regardless of the state of the lead
- If, apart from the lead, the roof appears to be in good condition then reconstruction with suitable substrates and chalk treatment might be sufficient
- Even if the underside of the existing lead is also in good condition, like-for-like replacement will not necessarily be immune from lead corrosion because the starting conditions may be different. Precautionary measures are desirable.

For most purposes when selecting substrates:

- avoid acid woods known to be chemically aggressive
- avoid fresh, damp and kiln-dried wood, or any with a pH less than 5.5
- avoid manufactured wood-based boards, particularly plywood, blockboard, chipboard, hardboard and oriented strand board
- keep the wood in a dry atmosphere for as long as possible before use
- do not use wood with a moisture content above 18%, unless the lead has been pretreated
- stop the wood getting wet during the laying process itself.

While penny (or wider) gaps have been traditional they do increase the amount of condensation when it occurs. In a carbonate-rich environment from chalk treatment close-boarding might possibly have advantages in some situations.

No underlay investigated yet has ideal characteristics. Geotextiles are good at providing air access and as a reservoir for chalk, but their high permeability to air also increases the amount of condensation under adverse conditions, and where laid over gapped boarding the chalk may fall out of the bottom.

To help to avoid initial corrosion, it is best to lay the lead in warm, dry weather: May to July is probably best. Conversely, in winter condensation is more likely. Humid, dewy autumn weather is virtually guaranteed to initiate some corrosion under fresh, clean lead, over gaps if nowhere else.

Over many years the experience of the LSA has been that most lead roof failures result from poor detailing, the most common of which are over-sizing and over-fixing (Coote 1994), leading to thermally-induced fatigue cracking. Where underside lead corrosion has seriously contributed to a roof failure, there has often been water ingress (leading to corrosion by the trapped moisture), thermally-induced cracking (which often starts where the lead has been weakened by the corrosion), or high concentrations of organic acids.

While some of the chemical processes discussed are exclusive to lead, underside corrosion failures should not be seen as specific to this material but as symptoms of underlying problems which will affect continuously-supported roofs of other metals to a greater or lesser extent. Moisture movement is virtually independent of the metal used for the roof covering and some other metals are susceptible to underside corrosion in unsuitable combinations of heat, moisture and chemicals. Similarly, any moisture problems identified potentially affect all types of roof, whatever the covering.

Future research

While many of the issues and problems have now been identified, in some ways this understanding has made solutions even more elusive than was first thought. In particular, some physical and chemical mechanisms which have assisted the survival of lead roofs in historic buildings in situations in which they are theoretically at risk are not yet well-characterized. These include:

- local buffering effects by moisture absorption in substrate timbers
- large-scale buffering effects of timbers and other hygroscopic materials in buildings and roof spaces
- self-passivation of lead in some roof spaces with limited ventilation
- dynamic heat, air and moisture movement around the lead/substrate interface.

We see the most important priorities as to:

- continue interpretation and analysis of the monitoring
- avoid including additional buildings in future tests unless there is good reason
- undertake further analysis of temperature, humidity and moisture content data
- test possible solutions to underside lead corrosion problems on site and in the laboratory
- investigate materials' properties and appropriate specifications for substrates and underlays

- consider appropriate details, taking into account the three-dimensional geometry of roofs and gutters.

THE UNDERSIDE CORROSION OF LEAD ROOFS IN HISTORIC BUILDINGS

1 Background

For centuries it has been accepted that lead roofs are susceptible to underside corrosion. Often this is cosmetic, though still undesirable as pollution by lead salts should be minimized. Occasionally, however, underside lead corrosion results in failure, sometimes by corroding through in places, but more often by concentrating thermally-induced stresses in areas thinned by corrosion, ultimately causing cracking and water ingress, which may then cause more corrosion: a cyclical effect.

Underside lead corrosion usually takes the form of a powdery, flaky white, pink or yellowish product, sometimes with traces of red and yellow lead oxides, particularly near the underlying lead's surface. While its chemical composition varies, basic lead carbonates usually predominate, though oxides, hydroxides, acetates and formates are often intermediate products which are then converted to the basic carbonate by the action of carbon dioxide in the air. Sometimes the carbonate is found (it is more likely to be formed initially in colder and drier conditions) and sometimes the oxide (more likely in warmer and damper conditions; P Forshaw, pers comm). Occasionally the corrosion product may be converted to sulphate, which is more protective. Deposits are seldom uniform, but in patterns which frequently relate to the geometry of the lead and of the underlying substrate, although in ways which sometimes vary surprisingly but are now beginning to be understood.

In the eighteenth century underside corrosion seems to have been regarded as part of the normal decay and renewal process of a long-lived material which nevertheless needed stripping and re-casting from time to time (Watson 1787). In the late 1970s, however, it re-emerged as a seemingly severe growing problem.

In the early 1980s the Ecclesiastical Architects and Surveyors Association (EASA) set up a sub-committee to investigate it. Its consultation document (EASA 1986a) concluded that:

- the natural durability of lead came from protective surface films which built up on exposure to the weather
- clean lead surfaces were readily attacked by pure water, usually in the form of condensation
- synthetic chemicals (such as wood preservatives) were not significantly involved
- organic acids from some timbers (notably oak, even if well-seasoned) when saturated could cause repeated aggressive action
- softwoods, unless degraded, did not cause significant underside lead corrosion
- the deterioration of lead in contact with new Portland cement was well-known. Although a separating layer of building paper was good common practice, it had a short life in damp conditions.

For lead roof design, the document suggested that:

- the combination of a clean underside lead surface and condensed moisture was sufficient to explain the corrosion observed
- occurrences of condensed moisture in roofs were likely to have been growing owing to changes in heating, ventilation and insulation
- pre-treatment of lead with a sulphate coating did not confer the expected resistance to sustained condensation (Hill 1982)
- the best principle was therefore 'no water, no corrosion'
- moisture under the lead should be avoided 'by ventilation or design'.

While the EASA report did not give firm recommendations, it discussed the three basic forms of roof: 'warm', 'cold' and 'inverted' (see also *Roofs and roof space environments*).

For 'warm' roofs the Building Research Establishment recommended a plywood deck, a vapour barrier of felt bedded in hot bitumen and insulation that was not moisture-sensitive and was capable of withstanding compression by light foot traffic. EASA was uncertain about the durability of the plywood, the rotting of timbers within the insulated zone and the best insulation to resist damage by puncturing and heat from the sun and from leadburning.

To avoid some of these problems EASA suggested that in a 'warm' roof one might consider a ventilated air space above the insulation and supporting the lead on a second plywood deck above that. This foresaw the 'ventilated warm roof' (see below), though not for the reasons it was finally adopted. Today one would not normally choose plywood as a decking, as discussed below in *Chemical properties*.

EASA were strongly in favour of 'cold' roofs but accepted their impossibility in buildings which did not have separate roof spaces. They suggested various methods of improving air flow, but current research indicates that these would not necessarily have been effective.

EASA also saw some merits in the 'inverted roof', where weighted-down insulation is placed over the lead, though in hindsight it is difficult to see why. Not only does this greatly increase the weight and change the appearance of the roof, but it also raises the water table (so the rolls and laps would no longer be watertight): the topside of the lead, no longer being exposed to the weather, might itself suffer from corrosion.

In 1986, just as the EASA report was being completed, news began to come in that 'warm' lead roofs were very efficient at trapping moisture. At first this was attributed to imperfect vapour control layers (VCLs), and this may sometimes have been true. However, contraction of the air enclosed between a good VCL and the lead as the temperature fell could create a partial vacuum under the lead which could draw in rainwater or moist air via rolls and laps by the so-called 'thermal pumping' process. In one notorious case (referred to in Murdoch 1987 and International Energy Agency 1994) a 'warm' roof built to the highest quality standards failed by this mechanism within four years, and after some tests warm roofs were no longer recommended (LDA 1988). These problems were advised to EASA members in an Addendum Sheet (EASA 1986b). To avoid thermal pumping, cold roofs (not permissible for flat roofs in Scotland) or ventilated warm roofs are now recommended (LSA 1993b, pp61–3; Murdoch 1987).

The above findings proved difficult for historic buildings. While one could often upgrade to a warm roof with little change in outward appearance using a minimal layer of insulation, ventilated warm roofs were quite another matter. It was not clear how effective 'cold' roof conditions could be attained in historic buildings, and ventilation could not entirely eliminate condensation, even in completely open roofs such as bell towers and Dutch barns. At the same time many roofs in historic buildings exhibited condensation but no significant underside corrosion.

In 1988 William Bordass Associates was commissioned for an *ad hoc* advisory consultancy to English Heritage on problems in historic buildings related to heating, ventilation and moisture. Recurrent questions included the appropriate environmental conditions for lead roofs and the difficulty of achieving the 'no moisture–no corrosion' principle in practice: nearly all roofs in historic buildings suffer from condensation from time to time. It was agreed to try to monitor when corrosion actually took place, to determine when environmental control measures would be beneficial and to test whether they worked. Following a meeting at the Society for the Protection of Ancient Buildings (SPAB), funds were made available for a research student at CAPCIS Ltd, the commercial wing of UMIST's Corrosion and Protection Centre, to adapt electronic equipment developed for continuous monitoring of steel corrosion for possible use with lead. After this proved successful in the laboratory, English Heritage funded a trial application at Manchester Cathedral (Dicken & Farrell 1990). The electronic monitoring also showed that corrosion was often fastest not when the lead was wet, but while it was drying out (Bordass, Dicken & Farrell 1989).

CAPCIS commented that such behaviour was not unusual in condensing environments (Farrell & Dicken 1990), owing to:

- a partial film of water on the surface promoting differential aeration cells
- evaporation of water causing trace elements in the condensate to concentrate and become more aggressive
- a faster corrosion reaction at the higher temperature.

The CAPCIS work and further *ad hoc* studies for English Heritage, the National Trust and SAS Software Ltd revealed an increasingly complex and often confusing situation. Some damp roofs exhibited very little underside lead corrosion, while some relatively dry ones had considerably more. Sometimes fresh lead samples corroded in roofs which had little or no corrosion, the pattern of corrosion under a freshly-laid sheet of lead could be very different from its predecessor and conditions at the time of laying could have major effects on subsequent corrosion behaviour.

In 1989 English Heritage identified the need for a more detailed research programme and over several years encouraged the BRE, the DTI and the then DoE to undertake work. Unfortunately, none of these initiatives bore fruit and in 1992 English Heritage decided to put together its own programme. The main work was undertaken by the corrosion engineers Rowan Technologies Ltd (RTL), with additional financial and technical support from the Historic Royal Palaces Agency and the Lead Sheet Association (LSA).

At the same time, William Bordass Associates was appointed to assist English Heritage with the technical management of RTL's project, to liaise with other research and with EH's advisory casework, to hold an annual forum Condensation Corrosion Forum of research workers in the field, see Appendix B, and to help to present the results in ways that were accessible to building professionals. The research is intended to:

- obtain a better understanding of underside lead corrosion and its avoidance
- investigate whether changes to the lead and its pre-treatment, to underlays and substrates or to the roof space environment can help to reduce underside corrosion
- consider improved specifications for lead roof repair and renewal, particularly in historic buildings where interventions need to be kept to the necessary minimum
- develop tools for corrosion diagnosis and risk assessment, to determine if a lead roof can be repaired or replaced much as it is, or whether it needs minor, or radical changes.

The research has been using a range of techniques to investigate underside corrosion and its avoidance, including:

- visits to sites with corrosion and those where the lead is performing well
- visits to sites where work has been done to attempt to avoid or reduce corrosion
- at some sites, monitoring existing conditions, in particular temperatures, relative humidities and timber moisture levels, and testing the corrosion behaviour of cleaned areas, various lead samples and remedial treatments
- constructing and operating two indoor laboratory test rigs which can each take eight samples of lead, with substrates and underlays where required, through a series of programmed test cycles of condensation and evaporation
- constructing, operating and monitoring outdoor test rigs with four different roof space environments: fully-ventilated (Dutch barn), ventilated air gap (ventilated warm roof), separate, partially-sealed roof space (as over a vault), and no roof space (roof as ceiling to internal space which was slightly humidified owing to a damp floor)

- experimenting with methods intended to reduce underside corrosion, and in particular different underlays, coatings and chemical treatments.

English Heritage has also been encouraging communication and joint identification of research needs between industry, research organisations, conservation bodies and professionals. Parallel studies of interest are also mentioned in this report, and are summarized in Appendix C.

2 Chemical properties

This brief review of the chemistry of lead and of the materials on which it may be placed picks up points which are particularly relevant to underside corrosion, arranged with the benefit of hindsight from the research to date, and including some results from the laboratory and field studies.

People are often surprised that lead corrodes. Chemically, however, the real surprise is that it does not, and some issues are more easily understood when seen in this light. Lead is attacked by distilled water, and more vigorously so in the presence of air, particularly when the water has a low carbon dioxide content (Hoffman & Maatsch 1970).

The main reason for lead's durability is that most of its salts are insoluble or sparingly soluble, the main exceptions being the moderately soluble oxide[2] and the highly soluble acetate and nitrate. If lead is left exposed to the weather, carbon dioxide and atmospheric pollutants, both from air and rainwater, react with it to form the protective grey patina which gives the material its traditional durability. While sometimes unprotective white salts are formed initially,[3] in due course these are washed off to be replaced by more permanent deposits which gradually take up sulphur[4] and oxygen and become increasingly protective (Tranter 1976). However, organic acid run-off from mosses and lichens can damage these, owing to the soluble lead salts formed.

Underneath, however, the lead is not necessarily passivated in the same manner and if clean lead encounters pure water in the form of condensation, underside corrosion may ensue.[5] Here the corrosion product stays in place and does not get washed off, so once corrosion starts here it is difficult to stop. In addition, any moisture trapped between the lead and the underlay, substrate or porous corrosion product may distil repeatedly, causing further damage, particularly where organic acids are present and also become trapped.[6] Poor access of air is also likely to make the trapped moisture deficient in carbon dioxide and produce corrosive and electrolytic effects.

A Building Research Bulletin in 1929 (Brady 1929) discussed the corrosion of lead by water, lime, cement, timber and soil. Key points reviewed include:

- Mortar can corrode lead owing to the combined action of moisture, oxygen and lime. This was attributed to the removal of any carbon dioxide in solution by interaction with the lime. Aged, carbonated lime did not have this corrosive effect. It was recommended that embedded lead and pipes should be either coated or wrapped, or otherwise packed with old mortar
- Serious corrosion was frequently caused where lead was in contact with timber, particularly oak, though softwoods could cause similar effects to a lesser degree. The presence of moisture was seen to be the controlling factor. Exclusion of moisture, and avoidance of wet and unseasoned timber, was recommended. Where poorly-seasoned oak could not be avoided, covering the boarding with bitumen felt was recommended
- In aggressive soils, it recommended wrapping pipes in bitumen felt, or bedding them in chalk, limestone or well-carbonated lime mortar.

While the above advice is still generally good, the current research has found corrosion on sites where oak has been covered with bitumen felt or bitumen-cored building paper. While these layers may keep most of the moisture and acids away from the lead, any small amounts that do get there cannot readily escape and can then do a disproportionate amount of damage. It also seems that chalk, carbonated mortar and limestone do not merely stop lead from corroding, but actively promote the formation of passive, protective layers.

Lead corrosion and passivation in aqueous environments
In dry environments lead does not normally corrode. In aqueous environments, lead (and all metals for that matter) may react with water, its components (hydrogen or hydroxyl ions), dissolved oxygen, dissolved carbon dioxide and the carbonate ion, or other dissolved materials (eg salts, air pollutants, organic acids and other carbonyl compounds). However, the lead does not necessarily corrode. Strategically, three very different outcomes are possible:

- immunity to corrosion: the metal can't dissolve and nothing present can react with it
- susceptibility to corrosion: the metal can dissolve and things can react with it. The reaction may or may not slow down depending upon the protection afforded to the metal by the corrosion product
- passivation: the metal can react with something present to form an insoluble product which may then protect the metal from further reaction. Strictly speaking, this should be called passivability because the insoluble product will not necessarily provide good protection: it may be porous, poorly-adherent mechanically or removed, by mechanical action or differential stresses under thermal cycling. For lead in air a curious mechanism also applies, see below.

The outcome will depend on many variables, in particular:

- the presence of all possible reagents, in solid, dissolved or gaseous form
- the possible chemical and electrochemical reactions between them

- the solubility of the metal and of all the possible compounds into which it might be converted in the above environment
- the stability (absolute and relative) of all the components that might be involved
- the electrochemical potential of the lead (expressed in volts)
- the acidity or alkalinity of the water (these are interdependent and normally expressed as pH)
- the morphology, porosity and coherence of the corrosion layer.

The susceptibility of a metal to corrosion over the full range of pH and electrochemical potential may be shown graphically on a Pourbaix Diagram. These diagrams are complex to construct as all possible equilibria have to be taken into account (Pourbaix 1966), but they are available in the literature for the lead/water system (Pourbaix et al 1966)

Figure 1 is the Pourbaix Diagram for lead in pure water. The dotted diagonal lines (a) and (b) represent the limits of stability of water itself: below the lower line it will be decomposed to hydrogen and above the upper line to oxygen. For lead on a roof, the main region of interest on the horizontal axis is between pH 3 (acid) to pH 12 (alkaline), and on the vertical axis at around zero electrode potential (E[V]), although locally, owing to the effects of differential aeration, potentials may vary within the bounds of the dotted diagonal lines. In this region of interest, there are no zones of passivation or immunity.

In the presence of dissolved carbon dioxide, however, Figure 2 tells a very different story. Now the insoluble carbonate (Cerussite – $PbCO_3$) provides a potentially safe bridge in the pH range 5 to 12 across the previously continuous corrosive band between the domains of immunity and passivation.

In practice, however, the corrosion behaviour of lead in the presence of air and moisture is more complex (Hoffman & Maatsch 1970, pp302–4). Partially immersed lead is strongly attacked by distilled water containing air, owing to the diffusion of oxygen and to concentration cells: the more deeply immersed the lead the less the corrosion.

The Pourbaix Diagram in Figure 2 was constructed at a partial pressure of carbon dioxide of one atmosphere. In outside air, with partial pressures of about 1/300th of this, Hydrocerussite (2 $PbCO_3$. $Pb(OH)_2$) is more stable than Cerussite ($PbCO_3$) at all pH levels (Edwards 1994), but because it is even less soluble, the general form of the Pourbaix Diagram is similar. Indeed, when the water is rich in carbon dioxide a protective film forms, and while it is not entirely effective corrosion proceeds only slowly.[7] Hoffman calls this Attack Type I (Hoffman & Maatsch 1970).

However, when condensation is fresh or when there is moisture in confined spaces (for example, between lead and an impervious underlay), the concentration of dissolved carbon dioxide is much lower. Now the air, carbon dioxide and insoluble Hydrocerussite join forces in a different and more aggressive process: Attack Type II. The mechanism goes like this:

The oxygen first dissolves in the water and begins to attack the lead: the dissolved carbon dioxide concentration at this stage is too low for Attack Type I to occur. The lead dissolves in the carbonate-free water, creating the hydroxide (lead oxide, PbO, may also be formed): $2Pb + 2H_2O + O_2 \rightarrow 2 Pb(OH)_2$. This diffuses to the surface where it reacts with atmospheric carbon dioxide to form insoluble Hydrocerussite: $3Pb(OH)_2 + 2CO_2 \rightarrow 2 PbCO_3.Pb(OH)_2 + 2H_2O$.

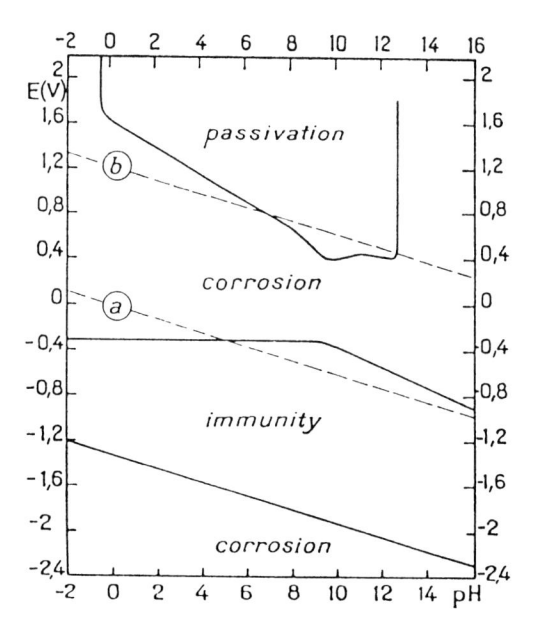

Figure 1. Pourbaix Diagrams showing domains of immunity, passivation and corrosion for lead in pure water (Pourbaix et al 1966, 485–92).

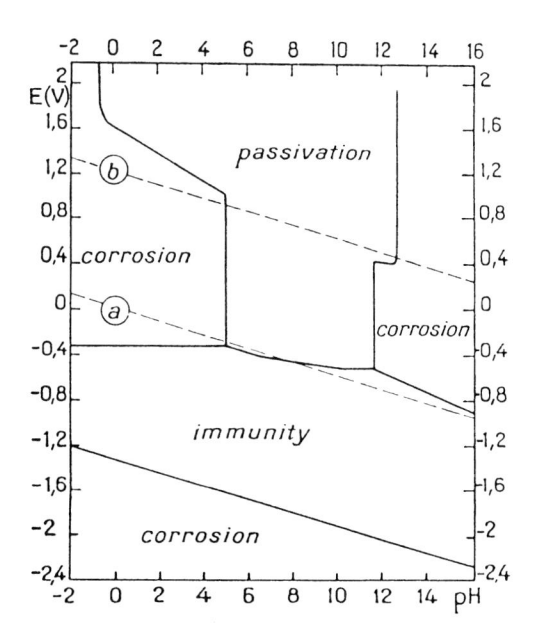

Figure 2. Pourbaix Diagrams showing domains of immunity, passivation and corrosion with added carbon dioxide (Pourbaix et al 1966, 485–92).

The above reaction both fixes the carbon dioxide which might otherwise have dissolved in the water and helped to passivate the lead, and also removes lead ions from solution allowing more lead to dissolve. In the absence of CO_2 the lead dissolves more slowly.

This Type II Attack is non-protective. From time to time the thin crust cracks and is repeatedly replaced. Loose crystals of lead oxide and hydroxide may also be found near the lead surface. Electron micrographs have been produced which show the porous arched non-protective structure of this crust (P Forshaw, pers comm) and on site one sometimes finds underside corrosion which looks superficially dry but which exudes moisture from its pores when scraped.

Passivation processes

If the lead surface cannot be kept entirely free of moisture, something which is normally impossible, it would be best to ensure that it is passivated and resistant to attack by moisture (whether or not air is present) and as far as possible resistant to attack by organic acids as well. This section outlines methods of passivation which have been attempted in the research to date.

Although neither Type I nor Type II corrosion achieve the passivation anticipated by Figure 2, under laboratory and site conditions effective corrosion-resistant passive films can sometimes form spontaneously beneath the lead. In RTL's laboratory condensation test rigs, passivation was observed (RTL 1993–5) under the following conditions:

- for samples exposed to the moist air of the test rig but not artificially cooled
- around the fringe of corroded areas of cooled samples exposed in the test rig.

In both circumstances it appeared that lead exposed to humid but non-condensing (or only lightly condensing) conditions could become passivated rapidly, probably because with only a thin film of water, sufficient carbon dioxide could reach the lead surface.

Since so much can depend on the initial surface condition of the lead, the research has looked into various ways of reliably protecting the underside. Pre-formed layers such as sulphate had already been investigated by EASA (1986a), with disappointing results. Simple, straightforward methods were therefore sought, using readily available safe materials.

- *Self-passivation in air.* Lead sheets usually come to site shiny but if left unrolled for a few days they begin to go dull owing to the formation of an oxide film (Hoffman & Maatsch 1970, p268). Vernon 1927 reported that in his laboratory environment this film was protective but on one occasion the lead instead corroded rapidly, while the pre-tarnished lead was unaffected: he attributed this to turpentine vapours from painting nearby. This film could well help avoid initial corrosion in marginal environments, but in RTL's test rig its resistance to repeated condensation cycles was relatively small in relation to weathered lead (see below)
- *Self-passivation when exposed to the weather.* After two to three months exposed to all the elements, a very corrosion-resistant film forms on the exposed side. However, the time required is unrealistic, the performance varies with location and weather and the problems of damage by working on site remain
- *Application of linseed oil.* Until the 1960s, linseed oil was often used in the rolling process. It persisted in some mills into the 1970s, and a few 'old' mills occasionally used for special purposes (such as sheets over one metre wide) still use it. It may have provided some protection, both directly and by forming lead soaps, which the newer, water-based lubricants do not. It has also been said (R Murdoch, pers comm) that plumbers used to have to scrape new lead clean before lead-burning (which is reportedly not necessary today) and that some of them used to wipe it down with linseed oil afterwards. In the present study, application of linseed oil has been found to confer some corrosion resistance, both in the laboratory and on site. However, there were practical difficulties such as the time it took to cure. If laid before curing, oil might be lost and moisture absorbed, leading to some corrosion. The sticky oil might also anchor the lead to the substrate and make it vulnerable to failure through restrained thermal movement
- *Application of patination oil.* This performed better than linseed oil and was more convenient, having a shorter curing time (typically overnight), though even this may slow down the roofing operation. However, we have found plumbers who pre-form their lead on jigs in a site workshop and apply patination oil to the underside at least a day before fixing, and claim that the time spent is not unreasonable, and that pre-forming brings some productivity gains in standardization and working in bad weather
- *Application of silicone spray (for instance, WD40).* This is sometimes used by conservators of lead sculpture but proved disappointing in initial tests. However, on one site, after one year treated and untreated areas were similarly corroded, but two years later there was less corrosion on the treated area
- *Application of silicone wax (for instance, Waxoyl).* Little benefit was found from this treatment, despite initial hopes
- In situ *chemical passivation.* The Pourbaix diagram and evidence from several sites where lead was found to be passivated over weathered concrete and cement-bonded woodwool slabs led RTL to investigate treatment using carbonate salts. Calcium carbonate (chalk and limestone), sodium carbonate and sodium bicarbonate were tested and finely-powdered chalk proved best (RTL 1995, Report 6). A slurry of finely-powdered chalk (3 microns average particle size) in water was applied by paintbrush to a coverage of some 200 grams of chalk per square metre (or sufficient to cover the lead surface). A durable, protective patina was obtained within ten minutes at room temperature.

While passivation, either in the test rig, on site or using chalk is visible as a darkening of the lead surface, the protective layer is thin and has not yet been well characterized chemically, in spite of exhaustive studies. It may be primarily an oxide [8] (as was also surmised by Vernon (1927), a basic carbonate or a combination of the two.

When lead becomes exposed to dampness, underside lead corrosion will tend to occur unless passivation has already taken place or the conditions are passivating, or preferably both. Lead roofs that perform well are usually either so dry that moist conditions are largely avoided, or have become passivated in the course of their lives and often, it seems, in the early stages. Once underside lead corrosion has begun to occur, although it will not always be serious, by then it is difficult if not impossible to revert to the highly-protective, thin passive layers which are sometimes found on site and can be generated in the laboratory. Of the various methods tried by RTL to promote the rapid formation of passive films, at present the chalk treatment looks particularly promising: not only is a good patina formed rapidly but if the chalk is left in place (for example spread over an underlay), then the environment may remain passivating, helping any failures to repair themselves. Preliminary results from site tests also indicate that chalk left in place may perform better.

Corrosion of lead by wood and wood-based products
Lead is susceptible to severe attack both by contact with damp wood and by the acid vapours emitted[9] (DoI 1979). Corrosion of lead by oak has been well known since historical times, being a method by which white lead was formed. Brady (1929) cautioned against laying lead over oak and over wet and unseasoned timber. Two more recent publications describe the corrosion of metals by wood in more detail (BRE 1985, DoI 1979). Corrosion of metals (and particularly lead) by acid vapours from wood without contact or condensation, in particular in the confined environment of a box, drawer or display cabinet, is well-known to museum conservators who refer to carbonyl pollution, by formaldehyde, formic acid, acetaldehyde and acetic acid (Grzywacz & Tennent 1994).

Organic acids naturally present in wood, including formic and particularly acetic acids, are the aggressive agents (Hoffman & Maatsch 1970, pp292–3) especially because the corrosion products, particularly the acetate, are highly soluble and so do not protect the surface. For acetic acid:

$2Pb$ [lead] $+ O_2$ [oxygen] $+ 4CH_3COOH$ [acetic acid] $\rightarrow 2(CH_3COO)_2Pb$ [lead acetate] $+ 2H_2O$ [water]

To add insult to injury, the effect is catalytic. Once formed, acetates (and formates) then react with carbon dioxide in the air to re-form carbonates or basic carbonates, regenerating the acid ready for a second attack, and so ad infinitum. Again for acetate:

$3(CH_3COO)_2Pb$ [lead acetate] $+ 4H_2O$ [water] $+ 2CO_2$ [carbon dioxide] $\rightarrow 2\ PbCO_3.Pb(OH)_2$ [Hydrocerussite] $+ 6CH_3COOH$ [acetic acid]

A small amount of trapped acid can therefore cause a disproportionate amount of corrosion. Over the years, there can even be cumulative growth in acetate concentration as more acetic acid is progressively absorbed. In sufficient concentrations (which site evidence for acetate suggests is over about 50–100 parts per million immediately below the lead), the acids may also attack protective films, such as the oxide and basic carbonate.

In general all timber species are acid to some extent, but some are more acid than others. Oak is notorious, while many of the common softwoods (such as white and yellow pine) are often both less aggressive and less prone to acid formation by chemical degradation. Table 1 lists various woods in order of their vapour and contact corrosion hazard. The woods in the 'severe' category tend to have pHs between 3 and 4, outside the 'safe' range in Figure 2, even before the acetic acid itself is taken into account! Those in the 'high' category are borderline, with pHs between 3.5 and 5.[10] As a general rule, hardwood tends to be more acid than softwood, and new wood more acid than old. However, acid is always present latently in wood (typically the acetyl group comprises between 2% and 5% of its dry weight) and is released in appropriate circumstances, for example by hydrolysis of acetyl groups in warm, damp conditions, for example in the kiln-drying process, during which acids (both originally present and from the hydrolysis) do not have time to disperse. Kiln-dried timbers can therefore be initially more acid than traditionally air-dried ones, though they contain less combined acid that could be set free in later years. In buildings this release continues for very many decades (DoI 1979) and may be re-activated even in old wood if conditions become damper (Werner 1987)

Manufactured timber products, including plywood, blockboard, chipboard, hardboard and oriented strand board can also be corrosive to lead,[11] as has been demonstrated both on site and in RTL's test rigs. There are four main reasons for this:

- the acidity of the constituent timber species, for example birch
- possible increases in acidity by hydrolysis during processing
- acids in glues and binders which often include or generate formaldehyde, formic acid (formaldehyde oxidises to this), acetic acid (viz polyvinyl acetate glue) and phenols
- a greater propensity to trap moisture.

Site evidence in this study suggests that underside lead corrosion can be very severe over moist wood-based boards, although one can find examples of dry boards

Table 1. Acid vapour corrosion hazards of woods (from DoI 1979)

Severe	Oak, Sweet Chestnut
High	Beech, Birch, Douglas Fir, Gaboon, Teak, Western Red Cedar
Moderate	Parana Pine, Spruce, Elm, African Mahogany, Walnut, Iroko, Ramin, Obeche

where all is well. On site, acetic acid also seems to concentrate in hardboard. At Donnington Castle hardboard was laid over oak (Lowe et al 1994, p6). Liverpool John Moores University found acetate concentrations of about 80 parts per million (ppm) in the oak and 600 ppm in the hardboard. Lead corrosion has also been observed in museum cases with hardboard, plywood, chipboard and blockboard (Oddy 1975), and the role of the adhesives was noted. In the museum world it has been concluded that in terms of corrosion risk wood-based boards have considerable disadvantages over carefully-selected natural timbers (Clarke & Longhurst 1961).[12] For lead roofs, the same advice appears to be equally appropriate.

Damp wood itself, whether acid or not, can also cause underside lead corrosion, by the processes already described. As a rule of thumb, if the moisture content at the top surface of the timber is:

- below about 15% it will tend to protect lead in contact with it from corrosion
- significantly over 20% it will tend to promote underside lead corrosion unless the acid content is low, the lead is well-passivated and evaporation-condensation cycles are infrequent
- between 15% and 20% or so it may sometimes promote passivation, if the acid content is not too high.

A Swedish study (Werner 1987) of the corrosion of iron found that:

- an increase in atmospheric relative humidity could cause acids to be released rapidly from wood
- the effect was much increased by raising the temperature by only a few degrees
- birch plywood and chipboard were particularly active, but 50 year-old pine, although much less active, also began to release some acid if atmospheric relative humidity and temperature were increased.

High humidity has a two-fold effect, both increasing the production of acid and the subsequent corrosion by that acid (Clarke & Longhurst 1961).

While EASA concluded that timber preservatives had no significant effect on underside corrosion (EASA 1986a), on site we have observed effects which differ for solvent-based and water-based materials. Further investigation is desirable. For the solvent-based materials:

- organic solvent-based preservatives can sometimes change the appearance of the underside of the lead and of any initial corrosion product. This may be a direct effect or possibly a leaching of resins from the timber. We have as yet no evidence of increased corrosion rates over the timbers, although on one site there was more corrosion over the gaps between preservative-treated than between untreated boards
- where bitumen-cored building papers are laid under the lead, the preservative's solvent has occasionally leached out the bitumen and brought it to the surface, sometimes adhering the paper to the lead, with possible adverse mechanical effects. Underside lead corrosion is often more severe over these leached areas, where moisture is more easily trapped.
- Solvents may also inhibit initial protection passivation (Vernon 1927).

For the water-based materials (both for timber preservation and for fire protection):

- often the timber comes to site very wet, initiating underside lead corrosion as soon as the lead is laid. Initial moisture content should be no more than 18%
- the preservation salts themselves may be hygroscopic, which will tend to increase the moisture content of the timber, at least when the atmospheric relative humidity is above the level at which the salts dissolve. Salts from sea spray or salt water immersion can have similar effects, as can the salt seasoning process, used in some parts of the world for controlling the drying of certain woods (DoI 1979)
- the salts themselves are often corrosive to steel and other metals: both chemically and because their constituent ions increase the conductivity of trapped and condensed moisture, and hence the rate of any electrolytic corrosion. This can often weaken nails and other fixings. However, for lead itself, salt contamination may slow down condensation corrosion attack because insoluble lead salts may be deposited (Hoffman and Maatsch 1970)
- EASA also noted that preservative salts could destroy building paper underlays, particularly those with aluminium foil facings (EASA 1986a)
- metal corrosion by preservative salts is also reported if timber is above 20% moisture content for long periods (International Energy Agency 1994, pp4–10).

From the work to date, it appears that the solvents, hygroscopic and conductivity effects may have more effect on underside lead corrosion than the treatment chemicals themselves. While EASA (1986) preferred solvent-based products, these are now criticized from the environmental point of view and water-based chemicals are becoming more common, as are their related problems. Many treatments (and certainly blanket treatments whether water- or solvent-based) are seen by environmentalists as unnecessary and needlessly polluting, particularly for decking which seldom seems to rot. However, indemnity insurance requirements and health and safety regulations are tending to force many specifiers towards the universal and sometimes unnecesary adoption of pre-treated timbers.

For avoiding corrosion by wood the following list has been developed from the Department of Industry's recommendations for packing cases (DoI 1979). For lead roofs, we have also added the points asterisked:

- avoid the woods in Table 1, particularly the severely and highly corrosive ones

- choose a wood with a pH value greater than 5.0 (any laboratory can determine this quickly and easily). For lead we would prefer a minimum pH of 5.5
- avoid, fresh, damp and kiln-dried wood
- avoid manufactured wood-based boards, and particularly plywood, blockboard, chipboard, hardboard and oriented strand board
- keep the wood in a dry atmosphere for as long as possible before use
- do not use wood with an initial moisture content above 18% (and preferably 15%)
- stop the wood getting wet during the laying process itself.

The Department of Industry also commented that to disperse acid vapours a small amount of ventilation was useless. Lime-washing the wood, which had been expected to absorb acid vapours, had also proved ineffective. However, in the current study laboratory and site tests suggest that the chalk treatment developed may provide some protection from organic acids, at least in the short term if in direct contact with the lead. The chalk used also had a pH of 8.9, a level close to that at which lead oxide and hydroxide is least soluble.

Chemical and metallurgical properties of the lead
Many building professionals feel that lead is not what it used to be, and are hoping for some new alloy which will resist underside corrosion. On the basis of the research to date, this seems unlikely. In accelerated tests in the laboratory, and in materials evaluation tests on site, no significant differences have yet been found between clean samples of sand-cast and milled lead, and between modern lead and cleaned lead taken from roofs up to 200 years old. Tests using continuously-cast (DM) lead are not yet complete. The main difference is in the surface condition: lead which for some reason has developed a passive film is the more corrosion-resistant, though in aggressive conditions even passive films from ancient roofs break down eventually.

The main reason for sand-cast lead's greater reputation for durability is probably its use in greater thicknesses than milled lead, so corrosion must be more advanced for visible failure to occur. The extra mechanical strength will also delay any onset of corrosion-assisted fatigue.

It is just possible that the 'steaming' of the underside during the sand-casting process might initiate a similar passivation process to that observed in moist but non-condensing conditions in the test rig (see above). As yet there is no direct evidence for this although RTL is now carrying out laboratory tests under milder conditions. Even if some effect were found, the information to date suggests that it would be unlikely to be substantial enough to be of more than marginal interest to a specifier.

In a small amount of work at Cambridge (Bordass, Charles & Farrell 1991) there was some indication that machine-cast material of high copper content (> 0.05%) with a coarse grain boundary distribution of the copper phase, corroded more quickly under water than milled lead, where the copper is more uniformly distributed. At lower copper contents there was no difference. One sample recently tested by RTL was also subject to pitting corrosion, but this may have been from a manufacturing fault. Further investigations are being undertaken and the tests repeated.

In wooden museum cabinets it has recently been shown that the amount of corrosion observed depends greatly upon the purity of the lead (Tennent, Tate & Cannon 1993). In one display case, a tin content of 1.5% rendered the lead resistant to corrosion. In another, containing a wide variety of lead badges, only two showed any corrosion: these were the only ones with a purity over 99%. While these alloys may well be inappropriate metallurgically for use on a building, and the conditions on a building will often be more aggressive, there may be something here worth investigating.

3 Roofs and roof space environments

This section discusses the construction of roofs and the underlying roof spaces, and the relationship of heat, air and moisture flows in these to the external climate and to the rest of the building. We start with requirements for modern buildings, as this helps to identify gaps between today's expectations and the actuality of historic buildings, many of which have performed well in practice. The difference partly relates to changed conditions in modern buildings: occupancy, management, habits, heating, ventilation, control, appliances, materials, insulation, materials and workmanship, which also creep up on historic buildings. However, some mechanisms which have helped roofs on historic buildings to survive appear not to have been fully appreciated, including moisture buffering by hygroscopic materials and conditions which may help to passivate the lead.

Roofing principles in modern buildings (Figure 3)
Roofs are normally designed on one of three principles (BRE 1986 and 1987), at least in theory:

- *WDI, warm deck roof: inverted construction.* Water-resistant insulation is placed on top of the weatherproofing layer. This design is not appropriate for lead roofs and so will not be discussed
- *CDR, cold deck roof.* Beneath the weatherproofing layer (lead in this instance) and its supporting substrate (and intervening underlay where fitted) is a space ventilated by 'cold' outside air. For flat and inclined roofs, this usually consists of a gap of typically between 25 to 250 mm in height, with ventilators at both ends. For ventilated pitched roofs (VPRs), except where there are rooms in the roof, there is often a walk-in or crawl-in ventilated roof space. Any insulation should occur below these ventilated spaces, usually with a vapour control layer (VCL) underneath it to control water vapour ingress, and often more importantly moist air ingress from the building, so we call it an air and vapour control layer (AVCL). Advanced designs may have a gap (sometimes insulation-filled[13]) between the AVCL and the internal lining to permit

services to be distributed without puncturing the AVCL. Condensation risks in CDRs increase with insulation level, imperfections in the vapour control layer and in cold or moist climates
- *WDS, warm deck roof: sandwich construction.* Here there is no outside air ventilation: there is insulation immediately under the substrate and an AVCL below that. This construction has been widely used for felt, asphalt, tiled and profiled metal roofs, with varying degrees of success although increasingly reliable products and specifications are now available.

For lead roofs, however, WDS construction, which was advocated by the BRE, the British Standards Institution and consequently by the LDA from the mid-1970s, created an unexpected and, for lead, a major problem. If the AVCL was poor, moisture from the building could come through and condense, creating a high risk of corrosion. If the AVCL was good, air trapped in the sandwich expanded and contracted with changing temperature, relieving pressure via the joints in the lead covering. In the contraction phase, 'thermal pumping' (International Energy Agency 1994, pp4–43) could draw moist air and occasionally even rainwater into the construction from outside. Occasionally movement under fluctuating wind gust pressures could do the same, if not for lead certainly for the lighter metals. Whatever its origin, the trapped water had severe consequences for the lead, especially if organic acids were present. While such water accumulation problems are not unique to lead (though splashlap details increase the risk) lead's corrosion chemistry makes its effect on the life of the roof particularly severe. In 1986 this construction was outlawed for lead roofs (LCA 1986) and this was subsequently recognised in BS 5250 (British Standards Institution 1989). However, survivors are still found, including a 1980 example which this study is examining.

To overcome problems with WDS, the 'ventilated warm roof' was introduced (LCA 1986). This has a ventilated air space between the upper surface of the insulation and the underside of the substrate. These roofs now tend to be constructed as fully-ventilated designs with ventilation openings at top and bottom, and so follow the same principles as the CDR.[14]

While there are no explicit UK recommendations for the size of the air gap in a metal-clad ventilated warm roof, for an inclined tiled or slated roof BS 5250 (British Standards Institution 1989) recommends the following:

- bottom inlet: equivalent to a 25 mm continuous slot
- top outlet: equivalent to a 5 mm continuous slot
- air path in between: 50 mm gap
- below obstructions such as rooflights: equivalent of a 5 mm continuous slot
- above obstructions such as rooflights: equivalent of a 10 mm continuous slot.

In practice, ventilated warm roofs found in the UK usually seem to have typical air gaps of 20–50 mm and inlets and outlets of 10–20 mm each.

Ventilated warm roofs are widely used in continental Europe for metal-clad roofs such as copper, zinc and stainless steel: lead roofs are not so common there. Recommended air gaps tend to be larger than in the UK. For example, the German specification based on DIN 4108 (Rheinzik GMBH 1988, 17–18)[15] recommends a 50 mm ventilation space for roofs above 20° pitch, rising to 100 mm between 3° and 20° and to 200 mm below 3° (and this includes beneath gutters!).

Recommended inlet and outlet openings are typically 1/400 of roof area (equal to 2.5 mm gap per metre length from inlet to outlet), though for pitched roofs the inlet size reduces to 1/500 (2 mm per metre).[16] Inlets should be as low as possible and outlets as high as possible, and if they are more than 10 m apart their sizes should be increased. For vertical cladding, the ventilation space is reduced to 20 mm and the inlets and outlets to 1 mm per metre.

Dutch practice for zinc-titanium is very similar to the German (Billiton Zink BV 1992). However:

- recommended ventilation inlets and outlets are of equal size
- for roof pitches over 20° the inlets and outlet sizes are reduced to 1 mm per metre
- underlays are not mentioned, instead sawn softwood boarding with 5 mm gaps is recommended, increasing to 10 mm if desired on roofs over 45° pitch.

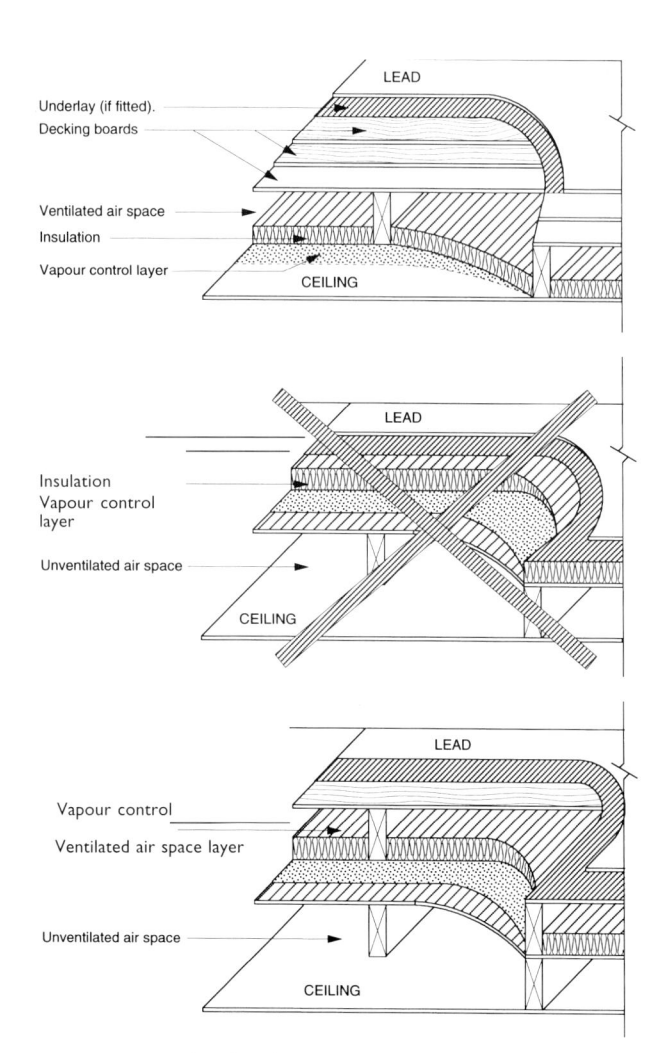

Figure 3. Principles of modern roof design: CDR (cold deck roof), WDS (warm deck sandwich) and VWR (ventilated warm roof).

For lead-clad stainless steel roofs the French (Centre Scientifique et Technique du Bâtiment 1991) require air gaps of 40 mm for distances up to 12 m and 50 mm for greater distances, but with only 0.3 mm per metre for air inlets and outlets, though with a minimum of 10 mm. Mountainous regions (over 900 m above sea level) require double-ventilated roofs, with a secondary weathering/vapour control layer between the two air spaces, of minimum 60 mm each.[17] Presumably:

- the lower ventilated layer allows any moisture penetrating from the building below to be ventilated away by outside air without condensing under the roof
- the upper layer allows the underside of the roof to be ventilated entirely by outside air
- the secondary weathering layer collects any condensation or melted ice that forms.

Complicated details are illustrated which maintain double-ventilation around obstacles.[18]

For the purpose of condensation control, BS 5250 (British Standards Institution 1989) identifies two types of 'cold' roof:

- roofs with limited space which may be difficult to ventilate adequately and where a vapour control layer would be necessary, for example a CDR (but equally a ventilated warm roof)
- roofs with a large ventilated roof space above the insulation, eg domestic pitched roofs in which vapour control layers are not normally required.

The IEA (International Energy Agency 1994, 4–11) states that it is very difficult to stop moisture moving into the roof cavity of a CDR, and the presence of water is very common. *Thermal Insulation: Avoiding Risks* notes the problems of underside condensation with metal-clad 'cold' roofs, and the importance of well-sealed vapour control layers. It regarded the CDR as 'a poor option in the temperate, humid UK climate' (BRE 1994).

BS 5250 (British Standards Institution 1989) also notes that:

- moisture should be extracted at source to reduce risk of vapour transfer to the roof
- water vapour penetration to the cold side of the roof construction should be minimized: constructional gaps and holes should be kept to a minimum and well-sealed
- vapour control layers should be adequately lapped and sealed, and their integrity maintained with puncturing avoided
- cross-ventilation openings should, if possible, be placed on the longer sides of the roof
- ventilation openings should be evenly spaced to avoid stagnant air pockets
- ventilation openings should be arranged so that they cannot be blocked, admit vermin or impair the weatherproofing. A 4 mm protective mesh is recommended
- the minimum free airspaces should be maintained past potential restrictions
- provision should be made to ensure moisture is not trapped and water vapour vented.

For pitched roofs:

- access doors should be heavy and clamped onto compressible seals
- high-level ventilation alone must not be used as it will suck moist air into the roof void
- materials which absorb condensate are preferred: they can re-evaporate it later when conditions are favourable. (However, some of this may end up under the lead.)

For both pitched and flat roofs, 25 mm ventilation gaps are recommended at the eaves along both long sides, while lean-tos should have a 5 mm gap at the top as well. The ventilation space should be at least 50 mm. For flat roofs over 5 m across, both the openings and the ventilation gap should be 'substantially increased' (though *Thermal Insulation: Avoiding Risks* [BRE 1994] suggests only a 20% increase for widths of 5 to 10 m). Attention is drawn to possible moisture attraction to thermal bridges where the insulation (and perhaps the vapour control layer) stops. This can easily happen at the eaves of refurbished buildings, and occasionally there have been outbreaks of timber decay after historic roofs have been upgraded with vapour control layers and insulation.

All the above recommendations are essentially based on the assumption that the roof is an inert structure which is subject to heat and moisture gains from below, and which must be protected from condensation damage by sufficient outside air ventilation to remove the moisture before it condenses. Recent research (for example, Cleary & Sherman 1987) casts doubt on this picture: there is a strong daily cycle in roof space dewpoint, which has been confirmed in RTL's studies, where on sunny days the dewpoint in roof spaces which are not generously ventilated can be well above those outside, and on cool nights well below. Such effects can potentially help to protect the lead: elevated daytime non-condensing dewpoints creating conditions more likely to provide passivation than in a generously-ventilated roof and the lower night-time ones avoiding some condensation events. There may also be seasonal effects: annual cycles in the moisture content of structural and decking timbers are well-known and the mass of water involved in these changes can be very large. For example if one tonne of wood increases its moisture content by 1% it absorbs 10 kg of water, which is sufficient to lower the relative humidity of 10,000 cubic metres of air at 10°C by over 10%.[19] A workshop on *Hygrothermal performance of the building fabric* at the Building Research Establishment on 21 January 1991 concluded that moisture movement and storage was poorly understood, current calculation methods did not necessarily describe the situation and further research was required. Subsequent work (for example, Jones 1993) has reinforced the importance of buffering effects and unventilated cold roofing systems are now being pursued in several countries (Shaw & Brown 1982, International Energy Agency 1994).

'Cold' roof design and condensation risks
Only the CDR, and its variant the ventilated warm roof, are now recommended for lead roofing. In essence they are variations on the same theme: a space under the lead and its immediate substrate, ventilated by outside air and from which water vapour and moist air from indoors are excluded by an AVCL. Paradoxically, Scottish Building Regulations prohibit CDRs but not ventilated warm or pitched roofs, probably because:

- for a CDR, with the ventilation normally within the structural zone, a good AVCL is very difficult to achieve and to maintain at ceiling level. People will always be putting in hatches and drilling holes for building services, which may easily destroy the action of the entire void. The zone itself is not easy to inspect and maintain in the way that a ventilated pitched roof is, and failures may threaten not only the weather-proofing but the structure
- for a ventilated warm roof, the vapour sealing, insulation and ventilation is restricted to a 'sandwich' on top of the structure. This permits a higher level of quality control: any failures will tend to be more localized, and if they occur they will tend to affect the roof finish and its supporting sandwich only, and not the primary structure. Vapour-permeable but waterproof sarkings may also be placed over the insulation to allow dispersal of any trapped moisture, and allow any condensation and water ingress that drips into the airspace to run out into the gutter
- ventilated pitched roofs, although in principle similar to the CDRs, perform more reliably in practice although their design has required some attention and improvement over the past two decades, as discussed below.

Condensation in roof spaces has tended to increase as a problem over the past 50 years owing to a number of mutually reinforcing trends:

- the change from coal fires to central heating has greatly reduced the amount of ventilation up chimneys, both hot and cold, and increased wintertime dewpoints
- the change from open-flued to room-sealed or electrical heating appliances, and from stoves to boilers in boiler houses has had a similar effect
- buildings without flues tend to carry more warm (and often moister) air into the roof space. Canadian tests on a two-storey house showed that a gas appliance with a chimney could reduce air leakage into the roof from the second storey of a house by 40% (Shaw & Brown 1982). Coal fires could easily have had a much greater effect and their sulphur-rich fumes may also have promoted passivation
- better sealed windows and doors tend to be opened less owing to changed habits, security requirements, noise control and energy saving
- more holes are made in the ceiling for building services (cables, pipes and ducts)
- heating is more controllable, and often operated intermittently
- in some buildings more internal moisture is generated, through increased tourism, catering, bathing and domestic appliances, and occasionally from humidifiers and unflued gas and oil heaters
- in some buildings lower occupancies, but with lower levels of heating and ventilation that can make them damper.
- progressively increasing insulation at ceiling level, which together with loss of incidental heating from chimneys in turn has tended progressively to reduce winter-time roof space temperatures and consequently increase fabric moisture levels
- former roof space ventilation paths are blocked, either deliberately or by insulation stuffed into eaves
- intermittent heating, particularly occasional heating of churches, can also increase roof space dampness: pulses of heat releasing pulses of extra moisture from the building fabric without creating enough of a continuous flow of heat and natural buoyancy ventilation to remove it (and the moisture from occupation) entirely (Bordass 1990).

When the present research began, it was felt that the above changes were the dominant causes of underside lead corrosion and that small alterations to heating and ventilation, perhaps only at critical times of the year, would make it possible to improve conditions and arrest corrosion. In practice, however, it appears that a 'golden age', where there was little or no underside lead corrosion, never existed. While the environmental changes outlined above have exacerbated the situation, they are contributory causes and not the sole reason, and adverse effects are difficult to reverse.

BRE Digest 270 (Building Research Establishment 1983) discusses condensation in insulated domestic roofs. It notes that in existing houses about 80% of the moisture enters the roof space with air rising from within the building, predominantly around roof hatches, via pipe routes and to a lesser extent through holes for electric cables and cracks at wall heads, and only 20% by diffusion through porous building materials. In a typical semi-detached house, 20-30% of the air dispersed via the loft, and in a single-aspect flat this became as much as 60%. Where chimneys were present, the proportions were reduced. Sealing hatches, access doors and holes and extracting air from moist spaces such as kitchens and bathrooms was found to be more effective at reducing moisture in roofs than adding a vapour control layer (which is both difficult to do retroactively, other than by paint, and as normally installed would tend to do little to reduce air leakage). To remove the moisture that did get through, roof space ventilation was recommended.

Current techniques for predicting condensation do not apply directly to constructions with ventilated air spaces. A recent laboratory study (Simpson, Castles & O'Connor 1992) of a flat timber 'cold' roof found that a vapour control layer was essential. It reduced the amount of moisture entering the roof space by a factor of 100.

With 0.01% of the polythene layer membrane's area perforated, the moisture gain increased by a factor of 10 but was still acceptable, but the authors noted that for specification purposes faultless sealing of the vapour control layer was imperative. Without this layer (but still with a plain plasterboard ceiling), moisture levels were high at distances more than 2 m from the air inlet, and in stagnant pockets across which the air did not pass. Under similar conditions, parts of a CDR tend to be damper than a ventilated pitched roof would be. This is because this type of roof contains a circulating body of air into which incoming air is blended while the CDR has more of a piston flow through the narrow gap.

In practice, ventilated air gaps in CDRs and ventilated warm roofs often fall short of good practice requirements. In particular:

- the ventilation does not run from bottom to top, as recommended, leaving dead spots above and below the ventilators
- sometimes other dead spots are not swept by the air path
- in complex geometries, and in particular at hips, valleys, rooflights, dormers etc, ventilation may be omitted entirely, or there is no through air-path
- occasionally gaps which should have been ventilated by outside air are instead ventilated, in whole or part, by inside air.

The UK's recommended air gap of 50 mm for a CDR (and implicitly for a ventilated warm roof) is smaller than that recommended by many other countries, at least for shallow roof pitches. To maintain the roofline in historic buildings contractors and architects have often been forced or tempted to use shallower gaps still. Unfortunately, however, the actual gap is sometimes even less than anticipated.

- In the test rig used by Simpson, Castles and O'Connor (1992), the researchers found that the nominally 100 mm-thick glass fibre quilt insulation had expanded to as much as 140 mm, so the designed 50 mm air gap was as little as 10 mm in places
- On ventilated warm roofs inspected we have found similar, though less severe, expansion. In one, a nominal 25 mm air gap became 20 mm owing to the effect of nominal timber sizes, and expansion of the 50 mm glass fibre batts had reduced this to 10-15 mm. RTL's studies indicate that 25 mm is probably a practical minimum (and then only allowable if the design and workmanship of the vapour control layer is extremely good)
- On several occasions we have also found air inlets and outlets wholly or partially blocked by projecting edges of the lead underlay or the vapour control layer.

Designers need either to allow extra space for fibre expansion (as is already noted in French-derived technical literature, see Eurocom 1993), to use more rigid insulation or to consider ways of stopping the insulation expanding (for example with a strong breather membrane on top and the air gap controlled by battens). [20]

Cold roofs are not the complete answer to 'no moisture, no corrosion'. Even if the building had an air and vapour control layer which was a perfect barrier to moisture migration into the roof space by diffusion and air movement, condensation is nevertheless likely in severe ambient conditions, particularly on still nights when radiation losses to clear skies takes the roof surface temperature well below ambient temperature and dew forms, or when a warm humid front arrives after a cold spell. Indeed, on dewy autumn mornings this condensation can often be observed on the underside of lead through gaps between the supporting boarding, but it is not necessarily accompanied by corrosion.

BRE Digest 270 (Building Research Establishment 1983) noted that while its recommendations suited most conventional roof constructions, under extreme overnight conditions more ventilation could actually increase condensation in lightweight sheeted roofs. However, it expressed the view that the extra ventilation would also clear the moisture more rapidly afterwards and prevent any long-term build up. However, this is not necessarily true if dripping or running condensation concentrates the build-up. For unpassivated lead, the resultant evaporation/condensation cycles can also be damaging, as already discussed in *Chemical properties* above.

Application to historic buildings
Few roofs in historic buildings comply with modern principles, except for very well-ventilated 'cold roof' situations, such as cloisters and bell-towers. Elsewhere, the underside of the decking often sees more air from inside the building than outside air. Even in nominally ventilated roof voids, the predominant air movement in winter (except on windy days) is often by natural buoyancy from below, with egress through both roof inlets and outlets.

Nevertheless, many such roofs have given good service, and their timbers do not get unreasonably damp. Modification to comply with CDR or ventilated warm roof principles is seldom easy, and often virtually impossible, taking into account technical, visual and historical constraints. Ventilated pitched roof principles may be easier to adopt, but where it is not possible to make good air seals between the roof space and the rooms below, additional roof-space ventilation may not only be unhelpful, it could be counter-productive, both for moisture levels (see Energy Design Update 1994) and in losing potentially useful buffering effects.

Roofs in historic buildings come in all shapes and sizes, but for convenience we have put them in four different categories (see Figure 4, A–D).

- Type A: direct onto occupied space, no intervening roof space
- Type B: underdrawn: a ceiling underneath but no distinct roof space
- Type C: domestic, with a roof space over the ceiling: this may or may not be ventilated
- Type D: stone vault, with a roof space above. A variant of C, but often more isolated from the building underneath and with more buffering capacity.

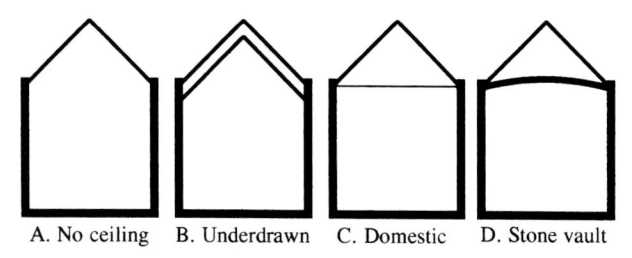

Figure 4. Four different types of roof.

A. No ceiling B. Underdrawn C. Domestic D. Stone vault

Many historic buildings, particularly churches, have Type A roofs with no ceilings, which fly in the face of today's practice and in which the underside of the lead and its supporting decking experience the same environment as the building's interior. In winter, the inside air will tend to be both warmer and damper than that outside: the extra warmth will tend to dry out the timbers while the extra moisture will have the opposite effect. On balance, interior timbers are normally drier than air-dried ones in protected ventilated spaces outdoors, as shown in Figure 5. However, in winter many poorly-heated churches have relative humidities in the region of 80% when unoccupied, so interior timber moisture contents can easily rise to 15% and more. Owing to heat loss to the outside, the decking timbers under the lead are colder and damper than this, and condensation risks are high. The processes are discussed further in *Influence on moisture levels*.

Underdrawn roofs, Type B, (known as 'cathedral roofs' in North America) with no explicit ventilation appear to offer the worst of both worlds: little or no opportunity to lower the dewpoint by introducing outside air (though a ventilated warm roof sandwich could potentially be added on top), while the winter-time temperature is reduced by heat loss to the outside, hence increasing the condensation risk (though this is only weakly dependent upon outside temperature, see *Influence on water levels*. However, these roofs seem to perform better than they deserve to, probably because the moisture-absorbing effects of the additional timbers between the lining and the roof decking can reduce the uptake of moisture into the structure and the amount of condensation, particularly under transient conditions.

Many domestic-type roof voids, Type C, act as an outlet for air from below, with some additional moisture diffusion for good measure. While additional ventilation is often called for, if it is not extremely generous it may be unproductive unless the sources or ingress are properly attended to first, which is usually much more easily said than done.

Roofs over stone vaults, Type D, are often found to be in good condition, probably for five main reasons:

- they are often on relatively dry, and continuously (if not generously) heated buildings, such as cathedrals
- they are often steeply-pitched, which reduces the amount of radiant heat loss under still, clear night sky conditions and helps water to drain from the splashlaps
- the steep pitch allows gap-boarding to be used, which although prone to condensation on dewy nights can dry out rapidly and does not trap moisture from any source

- the roof space is more separate from the air in the building (though access doors and holes can be a problem), with a greater proportion of ventilation by outside air
- they tend to be large in scale, and buffered by relatively large amounts of material.

The large amounts of hygroscopic material, especially timber in the roof spaces of many historic buildings, particularly Type D but to some extent also C and B, may have a variety of potentially beneficial effects both for the roof space environment and for the lead, and may work best with only limited amounts of ventilation by outside air:

- *Drying effects*: high roof void temperatures in summer may eventually dry out the timbers to a greater extent than if they were in free air, when they would not get as hot and where they would be able to re-absorb moisture more rapidly on cooler days and nights
- *Seasonal storage*: if the ventilation of the roof space is limited, the rate of moisture uptake by the dried timbers will be slowed down, allowing them to stay drier and to exert a dehumidifying effect during the vulnerably dewy autumn period. After the exceptionally hot summer of 1995, unusually dry roofing and decking timbers were found in many buildings right through the winter, which was also dryer than usual
- *Diurnal fluctuations*: when the roof is heated by the sun during the day, moisture driven off from the timbers humidifies the air and slowly escapes. When the timbers cool at night, moisture is absorbed and the roof is dehumidified. These augmented swings in dewpoint (William Bordass Associates 1986, Cleary and Sherman 1987) and have been reproduced in the unventilated roof void section of RTL's test facility
- *Improved passivation and protection of the lead*: the elevated (but non-condensing) dewpoints on sunny days may help to passivate the lead by a mechanism which would be less available in a better-ventilated roof from which the water vapour would disperse more rapidly. Similarly, the depressed night-time dewpoints may protect the lead, particularly above the gaps in the boarding, from transient condensation under some conditions.

4 Patterns of corrosion

This section reviews some observed patterns of underside corrosion and relates them to the circumstances in which they are found. For ease of description, the initial classification is by the type of substrates and underlays used. For each of these, other influences are mentioned, in particular:

- the geometry of the lead
- roof configuration and orientation
- the effects of external and internal climate (with more in *Influences on moisture levels* below)
- chemistry (see also *Chemical properties* above).

The issues are brought together more strategically in *Discussion* below. The information has been collected

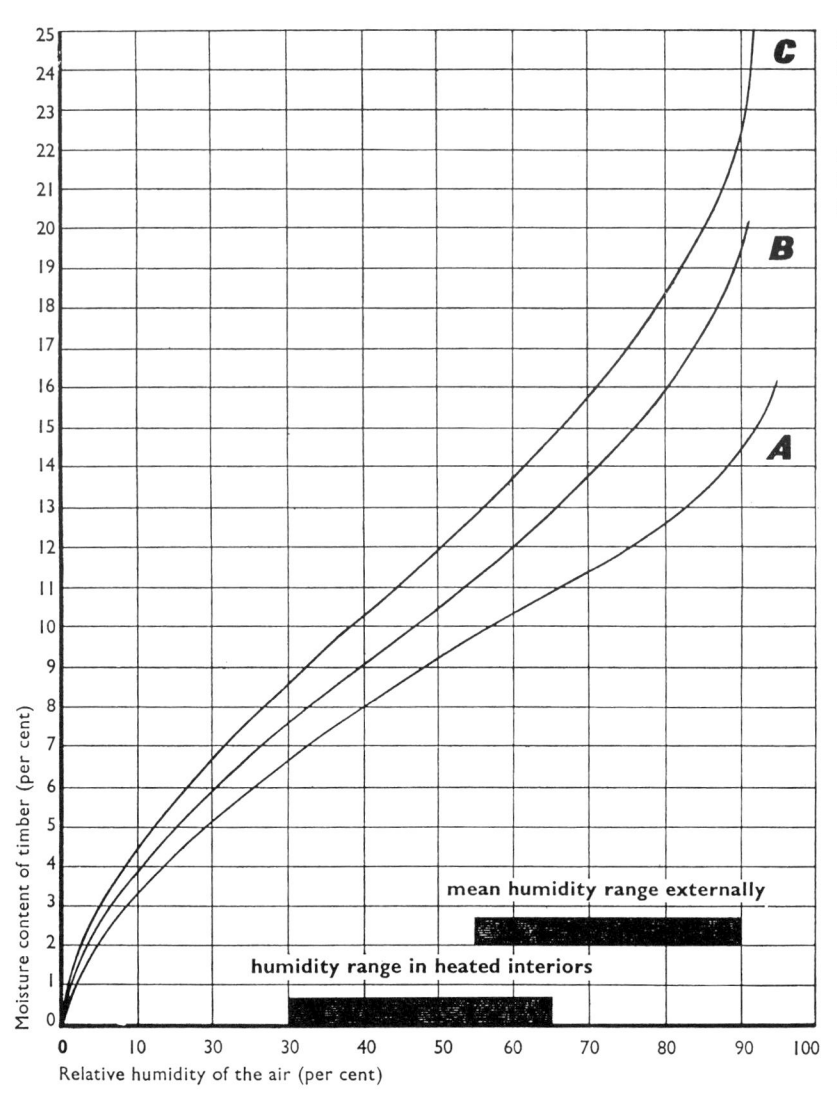

Figure 5. Relationships between equilibrium timber moisture content and indoor relative humidity (from Stillman and Eastwick-Field 1966, pp42–3). Many woods perform in the region of Curve B. Curve A includes teak, iroko, western red cedar and yellow pine. Curve C includes lime, sycamore and Corsican pine.

from nearly thirty sites visited, see Appendix D. Nearly two-thirds of these sites have received some monitoring and testing in this and associated studies. Most of the sites also contain a variety of roofs, of different pitch, construction and orientation, and sometimes with widely differing internal environments.

The underside surface of lead is often found in one (or usually several) of seven typical states:

- *State 1*. Type I corrosion (see *Chemical properties* above). A relatively hard, compact white scale, which confers some protection, unless it builds to such a thickness that it fails under mechanical stresses (often induced by thermal movement of the lead)
- *State 2*. Type II corrosion (see *Chemical properties* above). A looser, powdery or flocculent white scale. This tends to be formed in sustained condensing conditions. In more variable or gradually improving conditions it can sometimes 'harden up' and become more similar to Type I, although generally less protective
- *State 3*. Multi-layered flaky scales. Thick layers are often the consequence of prolonged Type II or Type I corrosion. While predominantly white, colouration by oxides may give them a yellowish tinge. Red oxide may also be found, particularly near the interface between the scale and the underlying lead. When they are particularly thick, organic acids are often present. Indeed, in most cases where corrosion has been sufficient to cause complete failure organic acids usually seem to be implicated.[21] Thin flakes are also sometimes found, probably the result of the roof being very damp to start with
- *State 4*. Localized streaks or spots of corrosion, often including brown as well as white areas. These are most often associated with the distillation of rainwater trapped in splashlaps into the adjacent roll or step. Corrosion only seems to occur in places where the two lead sheets are close enough together to trap condensate between them
- *State 5*. Dark and passivated. This tends to be found in environments which are sometimes moist but seldom fully condensing. These often occur in close proximity to highly corroded areas, even in acid-rich environments
- *State 6*. Dulled (sometimes with interference colours). This tends to be found either in dry environments where there is ample air (freshly-unrolled bright lead will dull down in a few weeks indoors), or in enclosed environments which are somewhat damper
- *State 7*. Bright and uncorroded. This normally occurs where the lead seldom if ever encounters moisture, for example when it is well-buffered by sound,

preferably low-acid, timbers in a dry environment. Usually access of air is also restricted.

Underside lead corrosion is seldom uniform, either on a single sheet or on different roofs over a building, and can vary tremendously between buildings, even where conditions are ostensibly similar. Many variables have both positive and negative effects: if the balance between them is slightly changed the outcome can be very different. If adverse physical effects occur (in particular where there is not only condensed or retained moisture but also frequent wetting and drying cycles), together with aggressive chemistry (in particular the release, generation and retention of organic acids), the results can be particularly severe.

Lead on close-boarded softwood deckings
Traditionally lead, particularly for low-pitched roofs, was often laid directly on sawn (or sometimes planed) softwood planks with 'penny gaps' between the boards.[22] While these gaps are prudent carpentry practice to allow for moisture movement of the timber, they also help to admit air to the underside of the lead, which in the right circumstances may assist passivation. However, the same route also allows moist air and water vapour to reach the underside of the lead rapidly, so increasing the risk and quantity of local condensation under adverse conditions, and providing a short-circuit path to the underside of the lead in contact with the boarding. On the other hand, the gaps also provide a path for any ingressed or condensed moisture to drip out, and for more rapid drying of the timber in the sunshine.

Corrosion under lead laid directly upon softwood boarding often reflects the patterns of boards, gaps and sometimes knots. Figure 6 is characteristic: on the nave roof of a church in Buckinghamshire, the lead is somewhat corroded (States 1 and 2) above the boards but above the gaps it is largely passivated (State 5). This passivation has occurred in spite of regular condensation above the gaps (at least in recent years). Freshly-cleaned lead samples placed above the gaps here also corrode.

However, as with the fresh lead samples mentioned above, one often finds more corrosion over the gaps than the boards, particularly for roofs laid in the autumn which can rapidly encounter condensation, if not from damp substrates, then merely from diurnal fluctuations in ambient conditions. For example, Figure 7 shows the ventilated warm roof section of the 1994–5 Donnington Castle tests in November 1994, about two months after these tests were started:

- over the top (leftmost) two boards of untreated softwood, which were laid in September 1994, the lead is only slightly dulled and shows interference colours (State 6)
- over the next two, preservative-treated, boards the lead is passivated (State 5)
- but at the joint between the preservative-treated boards there is a stripe of corrosion which widens out into the roll and at the top left licks into the uncorroded section.
- at the bottom (right) of the lower plank there is also some corrosion.

While the differences here may be influenced by chemicals in the timber preservative, evidence from computer modelling (see *Influences on moisture levels* below) and

Figure 6. Corrosion above softwood boarding. White and yellowish (States 1 and 2) corrosion product on the nave roof of a Buckinghamshire church, with dark passivation above the gaps between the boards. The photograph was taken after a cold, clear night, and the roof was particularly damp at the time. Note also the passivation over the rolls (with the odd spot of corrosion). The ridge is passivated on the far (right) side, probably by rainwater, which also seems to flow over the ridge from time to time (the lap is mean). Not surprisingly, there are signs of corrosion here, probably the result of a combination of condensation and refluxing of rainwater. In spite of this, and of the church being particularly damp (RTL 1993, Report 6, Appendix D), the corrosion is not yet serious. See also Colour Plate 2.

Figure 7. Lead after a two-month autumn test at Donnington Castle showing a wide range of surface states from passivation to corrosion. The two (untreated) boards on the left have kept the lead dry and its underside is only slightly dulled. The two to their right were preservative-treated and probably initially damper. This has assisted passivation over much of the boards, but corrosion around the perimeter. See also Colour Plate 3.

other sites suggests that initial moisture contents in the boards may be significant:

- the initial moisture content of the preservative-treated board may have been higher than in the plain board: here sufficient to assist passivation but not corrosion[23]
- when the lead was heated in the sun, moisture evaporated from the timber and passivated the lead more quickly
- the lead not in contact with the wood would not be passivated so well because the water vapour would disperse more rapidly

- however, this lead would be more at risk of condensation and corrosion when the temperature dropped (in the shade or at night), both from ambient moisture and from extra water vapour emission from the still-warm wood
- corrosion may have been exacerbated by volatile components of the preservative.

The church in Buckinghamshire also has an example of the high variability of corrosion patterns. White corrosion products are generally visible between the boards from inside the church below, while in the higher nave roof (above a clerestory) they are not (see Figure 6).[24] Figure 8

Figure 8. A church in Buckinghamshire: corrosion patterns on a slipped panel on the south aisle. See also Colour Plate 4.

shows a lifted sheet which is particularly informative because it has slipped at several stages during its life. Initially (when the lead was about one-third of a board width to the right of its current position) there seems to have been a strong tendency to corrosion over the gaps and passivation in between, but following the slippage some corroded areas have extended (as in the centre of the photograph) while to the right corrosion has been resisted by the passive layer. Indeed, it even appears that, in the lead's current position, one or two areas may actually have lost corrosion product and become more passivated by the present environment.

Lead on gap-boarded softwood deckings

Gap-boarding is commonly found in cathedrals and larger church buildings. The softwood supporting boards are often narrow (typically 50 to 75 mm wide, like tiling battens), with spaces of 15 to 25 mm between them. Since the lead has to span across the gap, they are most appropriate for the thicker codes of lead and for steeply-pitched roofs, for which there is also a possible advantage during construction as the sheets can be clipped over the battens at the top and intermediate clips used where necessary. They are usually installed without any underlays. Salisbury Cathedral used bitumen-cored building paper for a while but found that it made things worse.

Lead over gap-boarding in cathedral or similar roofs often seems to be in good condition, either passivated or with only small traces of corrosion (as at St Cross, see Figure 9; much of Salisbury Cathedral is similar), even though condensation can sometimes be found, for example on clear, still autumn nights. The effect is often attributed to good ventilation through the gaps alone, and certainly the avoidance of moisture traps and the ability to vent (or drip) away any excess moisture (and acids) will be helpful, shortening the time of exposure and potentially allowing passive layers to survive and regenerate. However, there appear to be other influencing factors, including gap-boarding's general associations with:

- more steeply-pitched roofs, which are not as subject to dew on still, clear nights because they cool less, owing to a diminished 'sky view'
- separate roof spaces, often over vaults, often well buffered and somewhat decoupled hygroscopically from the occupied parts of the building.

In turn the buffering may have other advantageous effects, particularly in more humid conditions when the roof is hot (which may enhance self-passivation), the ability of a buffered roof space to dehumidify itself after a warm period (which reduces the occurrence of condensation) and a slower transition into condensing conditions, which may assist passivation and absorption of carbon dioxide into the condensate (helping to avoid the unprotective Type II corrosion).

In an attempt to repeat the successes of past gap-boarded roofs, some new designs have reverted to gapped boards without underlays, for example in ventilated warm roofs. However, the results have been variable, probably because the buffering and passivating effects in a traditional situation are not necessarily present and the effects of any failure in the control of air and water vapour from the building below are more concentrated in the region of the failure.

For example, the foreground of Figure 10 shows a ventilated warm roof where the underside of the lead is in exemplary condition after two years. Above the softwood the metal is still bright, and above the gaps only dulled and perhaps slightly passivated. However, the bay immediately behind tells a different story of condensation and corrosion: not only above the gaps themselves, but spreading over the boards as well. The difference is that in the near bay the vapour control layer performed well, but in the adjacent one, and although the ventilated air space had been retained, faults in terminating the layer around the perimeter of a rooflight allowed moist air and water vapour from inside the building to leak into the ventilated airspace.[25]

This again emphasises the importance of meticulous attention to detail in design and workmanship of air and

Figure 9. Lead over gap-boards at the Church of St Cross, Winchester. This lead, which dates from the 1880s, is generally passivated above the gaps and very slightly corroded over the boards. As in Figure 8, there has also been some slippage. To get some corrosion in the lap is not uncommon.

Figure 10. Variations in corrosion performance over the gaps in a ventilated warm roof. Small amounts of leakage of water vapour and moist air can make large differences to corrosion in a roof which has limited buffering capacity. For comparison with later illustrations, note also the good condition of the rolls, which here do not have splashlaps.

Figure 11. Lead over building paper on softwood on a relatively dry building. The lead is bright or only slightly dulled, though with a white haze between the boards. The pattern in the lap is not unusual: at the bottom a weathered area indicating capillary rise of rainwater, some corrosion (with a fringe of dark passivation) above that (where the rainwater distils) and the lead above that unaffected. The hollow rolls are free from corrosion. See also Colour Plate 5.

vapour control layers if the intended performance is to be attained. The same applies to ventilation of the ventilated warm roof's air gaps: owing to its complex geometry, parts of the roof in Figure 10 were not fully ventilated and these also suffered from corrosion. On another building, the upper ventilators were not at the very top of a lean-to ventilated warm roof and underside corrosion was found in the region above them. Ideally, some protection to new lead in such situations would also be furnished, either chemically (perhaps using chalk treatment if this proves successful in tests) and/or using suitable underlays.

The effect of underlays on softwood substrates

Often lead is separated from softwood by underlays, which tend to be of three kinds:

- a building paper, often bitumen-cored
- natural felt blankets, made of animal ('hair felt') or vegetable fibres. These tend to be permeable to air and can absorb moisture. 'Inodorous' felt (eg Erskine's) (made of flax and jute with resin and pitch binders) was used frequently in the past but less so now because the binders become mobile at about 50°C (a temperature easily reached by lead in summer sunshine), which not only stiffens the material but may also stick it to the lead, which can then lead to thermal fatigue. If it rots, the decaying jute may also harm the lead
- polyester geotextile felts, which have become increasingly popular. These tend to be very permeable to air but the fibres do not absorb much moisture.

These underlays can affect the corrosion observed, as discussed below.

On a relatively dry building, building paper underlays can help to isolate the lead from the environment underneath, and avoid the patterned corrosion discussed above.

For example, the nave roof of a church in Northamptonshire has no void but just two layers of softwood boarding above (an older one underneath and a newer one on top). When re-leaded with cast lead in 1988, building paper was laid over the existing softwood. Two years later (see Figure 11), the lead was largely in bright condition, indicating little contact with either air or moisture. The softwood decking underneath was also relatively dry. However, at the time the church was unusual in being not only continuously heated but also in good condition and well cared-for by the churchwarden, who opened doors and windows on dry, sunny afternoons. All these measures helped it to be dryer than usual. Evidence of past condensation came from the architect's reports and from marks on the wooden ceiling underneath, suggesting that the situation might easily deteriorate if the heating and ventilating regime was to change.

In addition to the leaching discussed above, three main problems have been found with building papers:

- if bitumen-cored or polythene-backed building papers become wet on the upper side, from water ingress or condensation, the trapped moisture in the relative absence of air can exacerbate corrosion, particularly where acids are present. Even small amounts of water and acid may collect near defects such as laps and holes, causing severe localized corrosion damage and failure either by direct penetration or by increasing the lead's susceptibility to fatigue under thermal stresses
- condensation under the building paper cannot escape as readily under drying conditions, for example when the sun shines upon the lead, and so the risks of fungal or insect attack to the underlying timber may increase. The same applies to all unbreathable underlays unless they are carefully combined with additional insulation

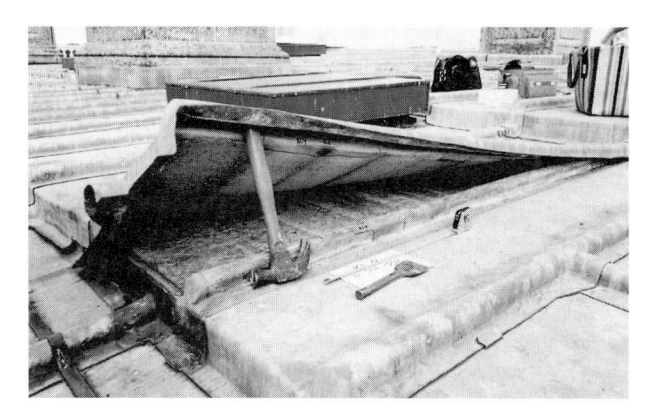

Figure 12. Underside of lead over Erskine's felt at a mansion in Dorset. The lead over the underlying boards above the public rooms is lightly and fairly uniformly corroded. Above the gaps there is a tendency to passivation. This corrosion is no longer active. Note also the corrosion in the roll above the splashlap caused by distillation of trapped rainwater: this recurs on a freshly wire-brushed surface. See also Colour Plate 6.

Figure 13. Underside of lead over Erskine's felt in a damper location at the Dorset mansion. The lead above the gaps is somewhat corroded, particularly towards the rolls in which there is also some corrosion. Note also the corrosion in the rolls above the splashlaps, as in Figure 12. The splashlap corrosion pattern varies with the distance apart of the sheets. Water is also trapped in the splashlap here by a slight backfall below the nosing. See also Colour Plate 7.

- the paper itself may rot, undermining its physical effect. Occasionally it has also provided a path for the rapid spread of fungus.

Natural felt underlays have three main effects:

- they provide some insulation, which in cold weather tends to make the wood warmer and the lead cooler. This can be significant in marginal cases (see *Influences on moisture levels* below).
- they allow air (and moisture) to permeate more uniformly to the underside of the lead, with mixed results
- they partially isolate the lead from the buffering effects of the wood underneath. However the hygroscopic fibres themselves also exert their own buffering effect.

Figure 12 shows some uniformly-corroded milled lead (laid in 1985) over Erskine's felt at a mansion in Dorset.

The corrosion product over the boards is thin, hard, compact and adherent Type I. Subsequent studies have shown that most of it was formed soon after the lead was laid, probably because the softwood decking (and perhaps the building) was relatively damp initially. In a few places where there was no Erskine's felt, the corrosion product was soft and flaky.

Over a warmer and somewhat more humid occupied flat in the same building, the lead above the boards shows similar Type 1 corrosion but over the gaps the corrosion product is whiter and less adherent, widening out into a 'fish-tail' as the roll is approached, and spreading into the roll and nosing (Figure 13). Wire-brushing patches (some can be seen in Figure 13) showed that lead freshly exposed in September was liable to further corrosion over the winter in the fish-tail areas only, where moist air rising from within the building could most easily con-

Figure 14. Corrosion over geotextile in a ventilated warm roof. While not serious, the underside lead corrosion is considerably greater than where the lead is placed directly on low-acidity softwood boards in ventilated warm roofs which are properly vapour-sealed from the building below.

Figure 15. Rapid initial corrosion at a mansion in Buckinghamshire. Four weeks after this sheet was replaced in the autumn, fish-tails were already very evident.

Figure 16. A year later in the same position the corrosion has now spread more widely.

dense. Corrosion did not occur in these positions (at least over a three-year test) in areas where the lead was cleaned and pre-treated with patination oil or linseed oil.

When geotextile underlays are used, corrosion above the gaps generally seems to be more pronounced in all but the driest environments. For example, Figure 14 shows some hazy corrosion where geotextile was used over a well vapour-sealed ventilated warm roof. This did not occur over plain softwood boards on other sites (see Figures 7 and 10).

Figure 15 shows a more dramatic example from a mansion in Buckinghamshire, where pronounced fish-tailing was evident under a replacement sheet only one month after laying in the autumn and continued thereafter (Figure 16). Interestingly this pattern is the reverse of that shown by the original sheets laid in 1905 over hair felt (Figure 17). Nevertheless, fresh (or freshly-cleaned) lead placed over the gaps corroded here too. However, lead laid in the early summer was much less affected as it had built up some passivation before encountering dewy winter conditions.

Three effects, all working together, appear sometimes to increase the likelihood of initial corrosion over geotextiles:

- the high permeability of the geotextile to air and water vapour
- the partial isolation of the lead from the moisture-buffering effect of the timber
- the non-absorbent nature of the geotextile fibres.

The effect of oak boarding

If any condensation or water ingress ever occurs (and even without liquid water), lead laid over oak boarding

Figure 17. The original roof at the mansion in Buckinghamshire. Most of the roof was laid over Erskine's felt and was passivated between and corroded above the boards (the opposite of the current pattern). A few sheets, which did not have the felt, showed a similar pattern but with worse corrosion over the boards, probably because any moisture accumulating within this region would be trapped for longer than over the more permeable felt, and possibly there might have been some acidity. The longitudinal stripe was made by wire-brushing for tests and is not part of the original pattern. See also Colour Plate 8.

Figure 18. Norwich Cathedral cloisters, showing lead distress and repairs. The pattern here is characteristic of acid attack, with not only weakening along the gaps between the boards, but also failure at the perimeter, at rolls and laps. See also Colour Plate 9.

tends to corrode much more extensively than in any of the examples yet shown, though nevertheless it may take 20 to 50 years before noticeable failure begins to occur. Figure 18 shows the pattern of repairs in the cloisters of Norwich Cathedral, and Figure 19 the very heavy corrosion underneath.

On several sites visited in mid-winter we have found the underside of lead laid over oak boards to be damp while over nearby softwood boards it is dry. Computer simulations by BRE Scottish Laboratory also predict this effect (see *Influences on moisture levels* below). The reasons for increased corrosion over oak may therefore be physical as well as chemical.

The literature suggests that lead may be protected from oak by interposing an impermeable layer such as bituminous felt.[26] On the sites studied we have found a variety of techniques, none of which has been entirely successful, at least for roofs directly over occupied spaces, with no intervening roof space.[27] Examples include:

- bitumen paper bedded in hot bitumen at church 1 in Yorkshire
- hardboard at Donnington Castle

Figure 19. Underside corrosion over oak at Norwich Cathedral cloisters. The patch removed, on the left, failed at the perforation in the middle. The characteristic red/brown oxide can be seen near the interface with the metal, with the thick, flaky and granular corrosion product remaining underneath (on the right). See also Colour Plate 10.

- hardboard over bitumen-cored building paper at the Great Hall, Hampton Court
- Erskine's felt over polythene-backed building paper at St Mary's, Stoke-by-Nayland.

In the first three buildings the protective measures have failed, typically after about 40 years, and replacement is now imminent. Stoke-by-Nayland, where the re-leading dates from 1967, is in much better condition but nevertheless shows some signs of distress. At a very damp church near Sheffield, lead laid on bitumen-cored building paper over timber boarding was also suffering acetic acid attack, probably arising from moist air and acetic acid vapour egress from the oak ceiling and the closely-spaced oak rafters underneath.

In simple terms, the failures could be attributed to 'an absence of ventilation under the lead', though as we have seen, ventilation by no means guarantees the absence of moisture or corrosion. However, all these sites share a relatively sealed environment between the lead and its substrate. Any water or acid which happens to get there by whatever mechanism will be trapped and available to hydrolyse timber products, dissolve adhesives and binders and distill with changes in the weather and sunshine. Even though barrier layers may restrict the total amount of underside corrosion, the remaining pockets of local corrosion can still be severe enough to require the whole sheet to be scrapped.

At church 1 in Yorkshire much of the underside of the lead surface is only lightly corroded, but there is severe corrosion, and penetration in places, where the lead rises into the hollow rolls and at the top of the laps (as in Figure 18). Essentially moisture and acid seem to have accumulated and distilled near the potential exit points from the roofing sheet. Here the atmosphere would also be richer in atmospheric carbon dioxide, to convert lead acetate into basic lead carbonate, and regenerate the acetic acid.

At the church near Sheffield moisture ingress was noticeable at the laps between the sheets of building paper, which were also uniformly wet. Since the laps ran down the slope of the roof, moisture also trickled down by gravity, as seen in Figures 20 and 21. Although the lead was corroded right through in places, and especially above gaps between some of the underlying boards (perhaps because condensation could collect here most), there were passivated areas nearby (dark shade), as in the lap in Figure 20 and in the frond-like pattern under the lead at the top right. The dark lines of the fronds coincide with the ridges in the puckered building paper underneath, suggesting that lead in close contact with the substrate was less corroded. These at first curious findings apply to many of the very badly-corroded roofs inspected:

- highly-corroded and well-passivated lead are often closely juxtaposed
- lead touching the substrate is often less corroded than where there is an air gap.

Donnington Castle gatehouse was one of the most corroded sites found. The roof of this unheated tower

Figure 20. Underside corrosion at the church near Sheffield. This roof over the NE chapel was very wet owing to a damp building, poor ventilation and a heating system which had been turned up, releasing more moisture into the atmosphere while the building slowly dried out. Some of this made its way through the construction and condensed under the lead where acetic acid was also present.

Figure 22. Typical underside corrosion patterns at Donnington Castle. A rafter underneath in the foreground coincides with the less-corroded area at the top.

was completely replaced in the mid-1950s by a new oak structure with oak rafters and purlins and an oak-boarded ceiling with hardboard over, presumably intended as a protective layer and to form a level surface for the lead. The lead began to fail in the late 1980s, and Figure 22 shows the heavily-corroded underside (average thickness loss 25%: (Lowe et al 1994), with patterning which coincides with the gaps between the underlying boarding (slightly less corroded above the gaps) and with the rafters and purlins (also less corroded over these). There is also some corrosion on the inside of the roll, but interestingly with a stripe of passivation at its base.

Across several bays near the southern eaves at Donnington there is a stripe of largely-passivated lead, see Figure 23. When the hardboard was taken up over this, a softwood fillet of approximately the same width as the stripe was found between two oak boards. There was also less corrosion above butt joints in the hardboard.

Chemical tests of the timber and hardboard showed that while the oak had a relatively low acetate content (for oak) of 80 ppm, acetate had accumulated in the hardboard (both original, from the oak and probably by hydrolysis) to a very high level of 600 ppm (Lowe et al 1994). Over the softwood strip, the hardboard's acetate content was only marginally smaller at 587 ppm, suggesting that the difference in corrosivity was likely to be primarily related to the physical (hygroscopic) properties of the softwood, and probably its higher absorbency. The

Figure 21. Valley gutter at the church near Sheffield. The lead from the gutter and roof has been removed in preparation for replacement by a ventilated warm roof. Products of corrosion at the oak rafter ends are very evident.

Figure 23. Passivated stripe at Donnington Castle. 3 mm hardboard separates the lead from the oak boarding underneath. The lead is badly corroded except for this dark passivated stripe, beneath which a softwood fillet was found under the hardboard. Since the acid content of the hardboard was found to be high (600 ppm) and similar throughout, the contrast is likely to be the consequence of different hygroscopic properties of the two woods.

Figure 24. The south-facing roof of the Great Hall at Hampton Court. Corrosion is greatest around the perimeter and in places where the lead has bulged away from the substrate. The corrosion in the rolls varies from place to place and from sheet to sheet, presumably depending upon the precise geometry and its propensity to trap moisture.

lighter corrosion over the rafters may also be an effect of their greater hygrothermal inertia.

From September 1993 to April 1994, RTL left a number of prepared lead samples sandwiched between the existing lead and the hardboard. The one over compacted fibre geotextile underlay was highly passivated. In contrast, at Hampton Court lead over geotextile was severely corroded, more so than over Erskine's felt. RTL's test rig work (RTL 1995, Report 6) also showed a reduced level of corrosion after the oak had become wet. As with pure concentration corrosion, there seems to be a fine, surprising and even more distinct boundary between passivation and severe corrosion when acetic acid is present.

The lead at the Great Hall of Hampton Court was relaid over hardboard in the 1950s, at much the same time as Donnington. On this much more steeply pitched roof, the hardboard has bitumen-cored building paper underneath. It too is heavily corroded, but the corrosion is more localized than at Donnington. Acetate in the

Figure 25. The north-facing roof of the Great Hall at Hampton Court. While this sheet generally exhibited more corrosion, it has a highly passivated area in the middle. Contrasts like this also occur on the south side.

corrosion product varied from 40–100 ppm in the lap to 450 ppm in the roll and 3700 ppm in the centre of the panel (ibid, Report 4, Appendix E).

Figure 24 shows a typical sheet on the south-facing slope. The corrosion is greatest around the perimeter, both just above the lap and beside and over the roll. Note also the tongue of corrosion a short distance above the lap, which coincides with a 'blister' formed in the lead sheet by restrained movement. Underside lead corrosion is also found under similar blisters further up the sheets (see also Figure 25). What these all have in common is that the lead seems to be worst corroded where it is least in contact with the acid and often moist substrate. There could be several contributory reasons:

- distillation of moisture and acid across the gap with changing weather and sunshine
- greater availability of carbon dioxide in the gap to regenerate acetic acid
- concentration of acetic acid during drying-out, although in the laboratory lead is less corroded by glacial acetic acid than dilute (Hoffman & Maatsch 1970)
- electrolytic effects (corrosion cells) creating anodic and cathodic areas.

Lead over plywood

In the recent past, plywood was often used as a substrate for lead. Today softwood boarding is again becoming more popular for all types of continuously-supported metal roofs, and the present study tends to support this choice. Plywood's performance can be acceptable if it is subject to little condensation or water ingress. However, if it becomes wet underside corrosion failure can be rapid, with severe symptoms occurring within typically 5 to 20 years. This seems to be the consequence of combined chemical and physical effects:

- the material is commonly made from acid timbers such as birch, Douglas Fir and some less well-known species, plus aggressive components such as formates in the glues
- the layered structure of the material, with vapour checks in the glue lines and generally poorer-quality and more porous timber in the middle lends itself to trapping moisture in the centre cores
- the trapping is exacerbated because the permeability of plywoods increases unusually quickly with moisture content, as illustrated in Figure 26. This means that in damp conditions plywood can take on large quantities of water, but if the surfaces of the plywood then dry out, water remaining in the core can be trapped, as we found at an educational building at Cambridge. BRE Scottish Laboratory are currently attempting to model this process
- any trapped moisture is then potentially available both to hydrolyse the core, releasing more aggressive chemicals, and also to move relatively quickly to the underside of the lead when the weather changes.

The flat-roofed extension of the conference/training room in a converted stables has lead on bitumen-cored building paper over a plywood deck, with glass fibre insulation underneath in the structural zone. The polythene vapour control layer beneath this is somewhat imperfect, owing to difficulties of making good seals to an inter-penetrating steel structure and at the perimeter. In a domestic situation the plywood would be at severe risk of condensation dampness. However, the occupation of the room is neither intensive nor dense, while its air-conditioning system provides good heating, ventilation and cooling when necessary but is not humidified. Consequently the internal moisture gains are relatively low.

Figure 27 shows one bay where the building paper was partly omitted. The pattern of light streaks and spots of corrosion is continuous across the edge of the building paper and so the lead may well have already been stained like this at the time of laying, or before. However, the background lead is darker in colour (and similar to fresh lead) above the building paper, while over the plywood there is a whiter haze. Corrosion on the outside of the roll may well be related to rainwater distillation from the splashlap, but the light corrosion at the inside of the roll

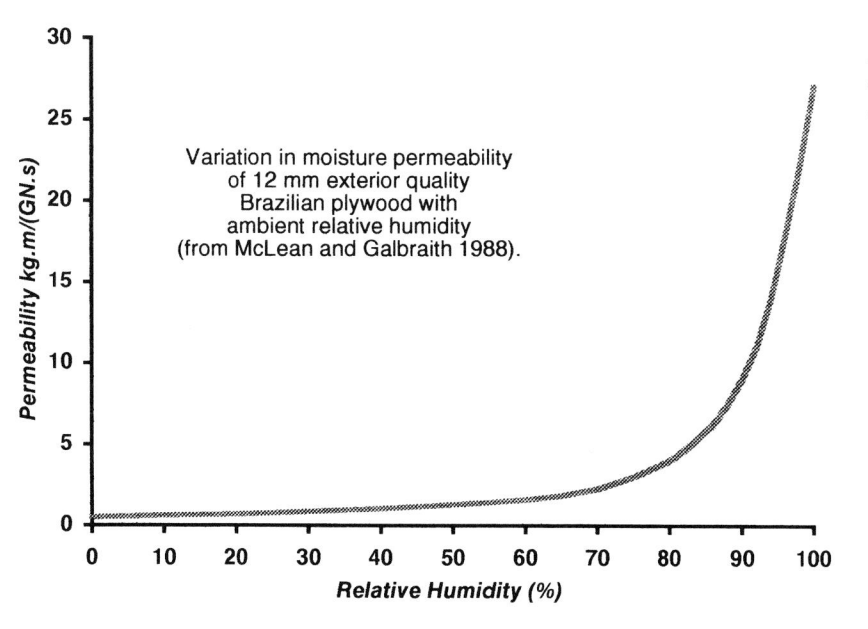

Figure 26. Graph illustrating relative humidity and permeability of plywood (McLean & Galbraith 1988).

Figure 27. Lead over plywood at a converted stables.

could be acid in origin, as might that above the step (which incidentally looks rather low). Further investigation is planned.

Figure 28 shows the underside of lead laid over a plywood gutter sole, where it was first subject to condensation corrosion and later to some water ingress. While badly corroded over the plywood, on both sides of this, where the substrate was woodwool cement slab, the lead is passivated. Theoretical calculations suggested that the lead over the woodwool would be more risk of condensation: while this was not physically evident, the architects for the remedial work reported that when removed it had lost some mechanical strength, a symptom of moisture-related deterioration. Owing to their open structure, these slabs may have carbonated rapidly and protected the lead by pH control and carbonate availability in the same way that chalk coatings have now been found to do.

Concluding points on corrosion patterns
This section has reviewed how different substrates and underlays can modify the effects of roof space environments and other variables upon the performance of the lead. Some key points are summarized in Table 2. Some characteristic patterns have been identified, including the fine line which often separates corrosion from passivation. Some things have been demonstrated unequivocally, such as the undesirability of acid substrates, including many wood-based panel products. Other things are highly equivocal: for example all underlays, and indeed ventilation itself, have both positive and negative effects which in practice often seem to be very finely balanced.

Figure 28. Lead over plywood in a gutter, with woodwool slabs to each side.

The underside corrosion of lead roofs in historic buildings 49

Table 2. Some physical influences on the underside corrosion of lead.

ITEM	POSITIVE EFFECTS	NEGATIVE EFFECTS	COMMENTS
ROOF SPACES			
1 Roof space ventilated to outside air	Can help to remove moist air and avoid condensation	May instead increase transfer of moist indoor air to outdoors via roof vents	May be of little or no net benefit if the roof space does not act as a theoretical cold roof and is poorly sealed to air and water vapour transfer from below
		Removes heat from the roof space and lowers its temperature	If there is still a high moisture flux from underneath, the roof may get little drier Near egress points, outlets and dead spots it can even be damper
	Conditions in the roof space are closer to those outdoors	Reduces hygrothermal stability of roof space	Hygrothermal buffering can enhance passivation, because the roof space becomes hotter and more humid (but not wet) when warmed by the sun
	Helps accumulated moisture to disperse when conditions allow	Introduces moisture with the outside air in dewy conditions	Hygrothermal buffering can also provide some protection against condensation in dewy conditions, with night-time dewpoints below ambient recorded on several sites
SUBSTRATES			
2 Underside of lead ventilated via air gaps	Introduces potentially passivating carbon dioxide	Carbon dioxide can enhance corrosion in some environments	Beneficial if condensation seldom occurs, particularly early in the life of the lead Once passivated, occasional later condensation may be tolerable
	Evaporates moisture quickly	Condensation is more rapid	Beneficial provided ventilation is 100% outside air; otherwise variable
		Faster evaporation/condensation cycles	Slight or occasional condensation may enhance passivation: heavier condensation may not Site location can be important
3 Lead laid on substrate of sheet material (no gaps)	Dry material helps to protect lead from condensation and corrosion	Wet sheet material promotes corrosion, particularly if acid, or becomes acid	Lead over wetter materials can perform well in relatively dry environments, but fails more rapidly in wetter ones, as effects of trapped moisture and acids combine
	Moist material may be passivating	But see above	Corroded and passivated areas are frequently found close together
UNDERLAYS			
4 Impervious underlays	Can protect lead from moisture and chemicals underneath	Can trap ingressed moisture and chemicals and partly exclude carbon dioxide	Can be very effective in relatively dry environments With water, vapour or acid present corrosion often severe, though sometimes localised
5 Highly air- and vapour-permeable fleece underlays	Introduces air and carbon dioxide and assists drying-out	Introduces more moisture in condensing conditions and lowers hygrothermal stability	More condensation, more evaporation/condensation cycles; and often more corrosion In marginal conditions corrosion is often worse if the underlay is non-absorbent
	Separates lead from potentially acid timber	Partly separates lead from buffering effect of timber, but acids may still diffuse across	Normally disadvantageous but can be useful reservoir for chalk powder, which not only enhances passivation but provides some hygroscopicity
6 Slightly permeable underlays (eg: plain building paper)	Provides buffer reservoir for overnight condensation	Not resistant to more sustained condensation and may occasionally rot	Useful to protect lead over well-ventilated and vapour sealed airspaces (efor example in ventilated warm roofs) against corrosion on dewy nights In some conditions, may also help to assist passivation
	Can disperse ingressed moisture	Reduces contact of air and carbon dioxide	On balance, seems beneficial in the circumstances described above
ROOF PITCH, ORIENTATION AND LOCATION			
7 Flat and low-pitched roofs	More consistently warmed by the sun, whatever the orientation	Get colder on clear, still nights, owing to radiation losses to night sky	As a rule more condensation, more retained moisture, more wet/dry cycling and more corrosion
		Roof spaces less easily ventilated	Moisture levels tend to be higher, generally or locally, increasing corrosion risk
		Any ponded water will cool the roof while it is evaporating	The outdoor test rig, with 100% outside air ventilation, showed corrosion under ponded areas, indicating how sensitive results can be to small variations
		Any corrosion product is vulnerable to damage by trampling	This can initiate spalling, open up cracks, and reveal fresh surface and traps for moisture, potentially increasing corrosion rates
8 Steeply-pitched roofs	Moisture can drain by gravity	This may sometimes be beneficial	Provided the water does not accumulate where it can do harm
	More likely to have roof spaces	Likely to be better ventilated or buffered	Normally beneficial
	Permits use of gap-boarding	Not good in humid environments	But can perform well in reasonably-ventilated and buffered ones
	South roofs may be drier	But may nevertheless be more corroded if the consequence is more wet/dry cycles	South roof performance tends to be the most variable
	North roofs may be damper	But not necessarily more corroded as there may well be fewer wet/dry cycles	Where acids are present, variations between different orientations often seem to be smaller (cycling may be less important than average conditions)
	Better air circulation in roof space	Sometimes moister near apex	Better buffering and more uniform conditions generally seem to be beneficial
9 Relative position	Lower roofs act as air inlets	Upper roofs act as outlets	Upper roofs are sometimes more corroded, but depends on heat distribution too
	Gutter area acts as air inlet	Upper areas act as outlets (but they are not necessarily more corroded as hot weather passivation can be greater too)	Gutters usually corrode less than one might expect, unless they trap moisture (e.g: running down from above) or are laid on acid timber or fresh concrete or lime Gutters may also suffer less because they are less exposed
LEAD DETAILING			
10 Splashlaps	Help reduce uplift and rainwater ingress in exposed positions	Trap rainwater (particularly at steps and near the bottoms of low-pitched sections) and inhibit drying of wet substrates	Promote thermal pumping and distillation at rolls & steps, particularly if tightly-dressed Corrosion may occur in the outer part of the roll May also inhibit drying of wet substrates, increasing corrosion on the inside of the roll

The illustrations have also shown the variability in patterns of corrosion at rolls, laps and steps. Some of this, particularly corrosion near splashlaps on low pitches, is related not to the internal environment or to substrate conditions but to the distillation of rainwater. Where distillation from splashlaps occurs, the corrosion on both the over- and under-cloaks usually matches and tends to occur in streaks or patches where the two sheets are sufficiently close together to trap condensate between them. Often the corrosion tends to be only cosmetic: where removed, it can re-establish itself quite rapidly but then settles down, perhaps because once the corrosion product forces the sheets apart distillate is less easily trapped. However, more such corrosion can occur near the roll-end nosings in low-pitched roofs, which sometimes need repair even where the rest of the roof is in good condition. In these positions the reservoir of water lasts for longer (particularly if there is a backfall), the water distilled into the nosing has nowhere else to go, and of course the nosings will often have already been thinned or otherwise weakened by bossing. However, on one site we have recently found more extensive splashlap corrosion, for reasons which are not yet clear.

Where moist air emerges from the buildings through gaps in boards, characteristic 'fish tails' are often seen, rising into the rolls. In new roofs, these are more often corroded: in older roofs areas of passivation are also found, though quite frequently with corrosion above the boards themselves.[28] Site investigations and RTL's test rigs also indicate that such different behaviour can be triggered by very small changes in environmental conditions. With the more permeable underlays (such as geotextiles) corroded fish-tailing increases, probably because with less resistance to the passage of air through the gaps between the boards, more condensation can occur under adverse conditions.

However, the heaviest corrosion in roll, lap and step positions is usually found where the lead is laid over sheet materials and over acid substrates such as oak and plywood. Here corrosion is often greatest at the perimeter, probably because trapped moisture, often by this stage containing acetates etc, will distill repeatedly before it finally emerges, in an atmosphere in which acetic acid can be regenerated rapidly. Acid-induced underside lead corrosion also tends to be worst where there are splashlaps, which will tend to inhibit the release of moisture, and sometimes provide some water vapour inputs of their own.

5 Influences of moisture levels

Moisture, particularly condensed and trapped moisture containing organic acids, is the principal agent in the corrosion of lead. The previous section has shown how sensitive lead can be to even small additional amounts of moisture, for example

- at the gaps between the decking boards
- in the parts of a well-ventilated 'cold' roof in which the lead surface is sometimes slightly cooler owing to evaporative cooling from puddles above
- in sections of a ventilated warm roof which are not fully through-ventilated or in which there are small imperfections in the vapour control layer under the insulation
- with a change in underlay.

On the other hand, one can also find lead in good condition and well-passivated in some situations which are manifestly damp.

This section outlines some mechanisms of evaporation, condensation and buffering which may affect the state of the lead and the underlying substrate, and how they may differ in different buildings with different construction, different levels of heating, ventilation and moisture generation, and at different times of the year. This work is not complete: more analysis is needed of the temperature and relative humidity data collected. There is limited availability of theoretical techniques which can describe the situations of interest, and little available information which could be used to calibrate them. Further investigation and analysis is strongly recommended.

The external environment

Figure 29 is a psychrometric chart of water vapour pressure versus air temperature. The darker line at the top, marked 100% RH, is the saturation curve, showing the maximum vapour pressure that can be sustained in air at the appropriate temperature: air can contain no more moisture than this (unless in a fog of dispersed droplets). The family of curves 80%, 60%, 40% and 20% RH shows a series of relative humidity levels, where the moisture content of the air is the given percentage of the saturation value. Figure 29 also includes the average vapour pressure and temperature for each of the 52 weeks of a typical year, using the most representative months from Kew Meteorological Office data for 1959–68 compiled by the Polytechnic of Central London (Loxsom 1986). In examining Figure 29, the following points are of interest:

- in general, the warmer the weather, the higher the vapour pressure, so air in the summer typically contains a lot more moisture by weight than in the winter. At a typical winter time vapour pressure of 650 Pascals (Pa)[29], air contains about 0.4% by weight of water, in summer 1300 Pa, twice this much
- in spite of this, in the warmer months the RH is lower because the air's moisture content is a relatively smaller proportion of the saturation value
- hence, in sheltered outdoor conditions, materials tend to be drier in summer. For example, using Curve B in Figure 5, timber in equilibrium with air at typical summer RH of 65% would reach a moisture content of about 13% and in winter at 85% RH, about 17%. However, it can take a long time for equilibrium to be reached, particularly for large, dense timbers in poorly-ventilated spaces. On the other hand, summer heating by the sun can cause extra drying, with structural timbers in warm roof spaces falling to perhaps 10–12%, and boards immediately under the lead to

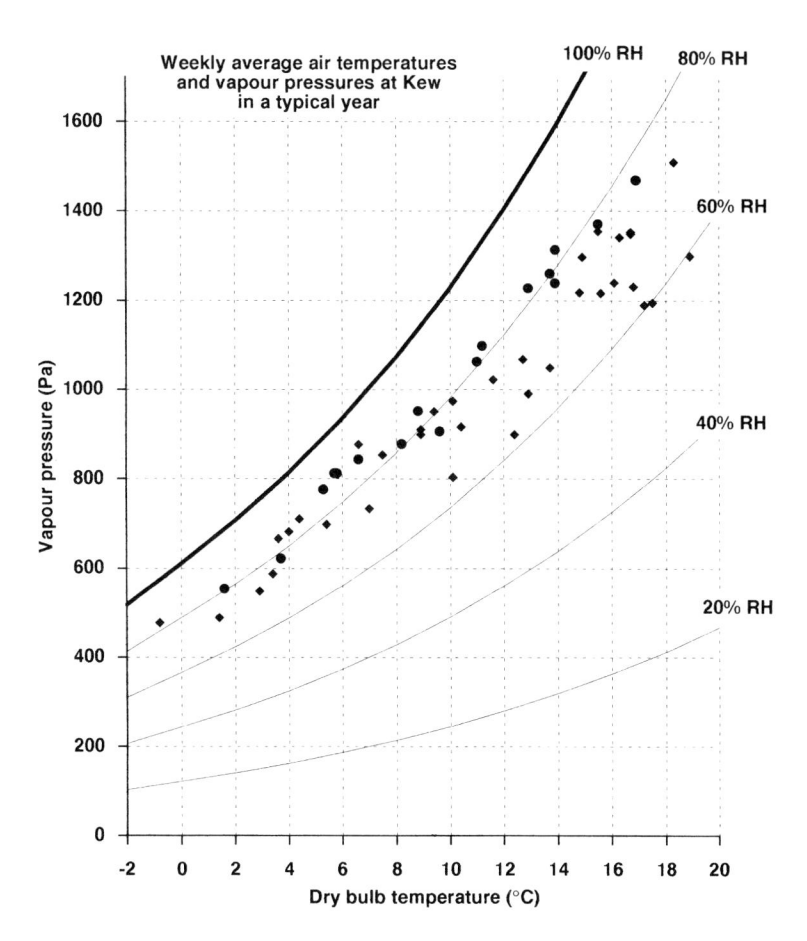

Figure 29. Psychrometric chart showing weekly average outdoor temperatures and vapour pressures. The points marked as diamonds show conditions in the months January to August while the solid circles show those from September to December, when the relative humidities tend to be higher, as is the likelihood of condensation arising from fluctuations in external temperatures and from further cooling of the roof by radiation heat losses on still, clear nights.

around 8%. In winter, some timbers can get damper, as discussed further below.

The points for each week are shown as diamonds for the months January to August and as dots for September to December: the four months in which RHs tend to be at the high end of the range. The high RHs in autumn arise essentially because the ocean has warmed up during the summer and stays relatively warm into the early part of the winter, humidifying the air which is carried over the country by the prevailing south-west winds. In this season, condensation arising from fluctuations in ambient temperature and from radiant heat losses from roofs on clear, still nights tends to be the most common: fresh, unpatinated lead can be particularly vulnerable to corrosion then.

The indoor environment
On average, and particularly during the heating season, the indoor environment will tend to have both a higher temperature and a higher vapour pressure than that outside. Even where there is no explicit heating, trapping of solar heat and heat from occupancy, lights and equipment tends to raise the internal temperature. Moisture is also added by the metabolism of occupants and often plants, by activities and equipment, in particular cooking, washing and bathing, and by evaporation from the building itself, for example from any rising or penetrating damp. Heating itself may also humidify the air through three main mechanisms:

- by warming up the fabric the rate of evaporation is increased. While in a dry building this moisture will eventually leave, if there are semi-infinite sources of moisture (for example from the ground outside), it may be continuously replaced
- if the building is heated intermittently, moisture will be given off by the fabric and contents into the air as it warms, giving additional transient humidification
- flueless gas and oil heaters also humidify the space directly by emitting their products of combustion, which include a lot of water vapour, directly into it.

Figure 30 shows a section of the psychrometric chart, with points plotted on it for the typical test situation as used in BS 5250:1989, the British Standard Code of Practice for the control of condensation in buildings for 'dry/moist' occupancy (British Standards Institution 1989). The RH curves here are plotted at 10% intervals. The outside air is at a typical wintertime state of 5°C 95% RH (in Figure 29 temperatures below 5°C occurred for nine weeks) and the inside air averages 15°C 65% RH.

If the outside air is cooled somewhat, its condition travels to the left along the horizontal arrow until it hits the saturation curve at its dewpoint of 4.25°C (see drop arrow). Any exposed surface any colder than this will attract condensation even from the ambient air: this can happen on clear nights (and indeed on bright winter mornings for lead not directly in the sun) when the lead loses heat by radiation to the clear sky. On clear nights, the lead can fall 5°C or more below outside air temperature, particularly for flat roofs, over a cold building or a well-insulated roof void, and in still conditions where the lead picks up less heat from the outside air.

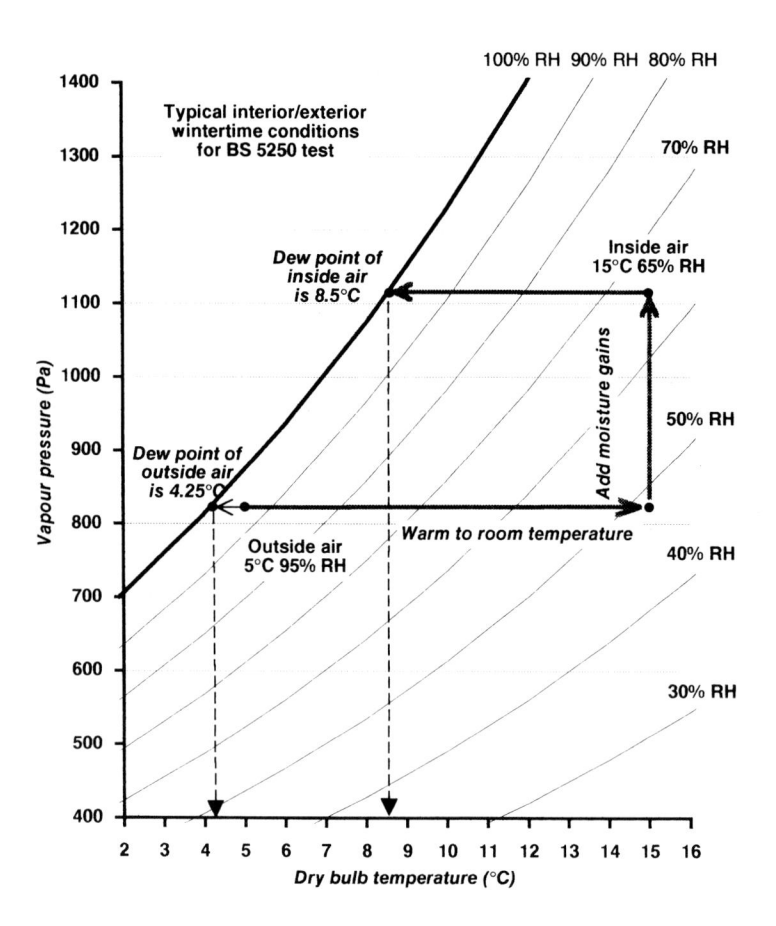

Figure 30. Psychrometric chart showing BS 5250 test conditions. External conditions are 5°C 95% RH (dewpoint 4.25°C). Internal conditions for 'dry/moist' occupancy are 15°C 65% RH (dewpoint 8.5°C), a vapour pressure gain of 285 Pa.

If outside air coming into the building is warmed to 15°C, its condition travels to the right along the horizontal shaded arrow, maintaining the same moisture content and vapour pressure but reducing in RH because the saturated vapour pressure of the warmer air is higher. In this example the RH falls to 48%. However, moisture gains within the building add to the vapour pressure, increasing it by some 300 Pa in this example, and raising the RH to the 65% test condition. The actual amount of water vapour pressure gain in practice will depend upon:

- the rate of evaporation of moisture into the air by the mechanisms outlined above
- the ventilation rate of the building, in the course of which more moist inside air is displaced by drier outside air (in terms of percentage by weight)
- (to a lesser extent) diffusion of moisture through the fabric of the building.

With high rates of ventilation or low moisture gains, the inside air will be drier, with low ventilation rates and high moisture gains it will be damper.

Dampness levels in a building, and in different parts of it, can vary greatly with heating, ventilation and construction. For example:

- with generous heating and ventilation, the horizontal displacement to lower RHs is much greater than the vertical displacement by added moisture, leading to a drying effect. For example, at 15°C 48% RH caused by the heated and unhumidified outdoor air in Figure 31, the equilibrium timber moisture content would be about 10%, and if the air were further heated to living-room temperatures of 20°C the RH would be in the low 30s and timber moisture content 8%. Again, these are limiting values subject to time lags (frequently measured in weeks or months) as conditions change
- with generous heating but with poor ventilation, moisture may build up internally, as in the example in Figure 31. If this moist air then rises into the roof space, it may then cool, say to a temperature of 10°C, and if not diluted by outside air its RH would approach 90%, when the equilibrium timber moisture content would be nearly 20%. With a dewpoint of 8.5°C, condensation is also likely under the lead
- with poor heating and ventilation, the atmosphere will be cold and damp, particularly if structural dampness adds additional moisture. With occasional heating, as in churches, the pulse of heat evaporates some of this moisture and raises the dewpoint further.

For a given amount of heat and moisture input, there is an optimum amount of ventilation to minimize RH. As the ventilation rate increases from zero, the RH first drops sharply as moist air is displaced from the building or roof space. However, with too high a ventilation rate the heat is removed too rapidly, the temperature falls and the RH rises again: to reduce RH further either requires more heat, less moisture input or moisture removal at source. In domestic circumstances, optimum ventilation rates tend to be between 0.5 and 1 air changes per hour: beyond this additional ventilation is usually counterproductive because the additional heating required is seldom regarded as necessary or affordable.

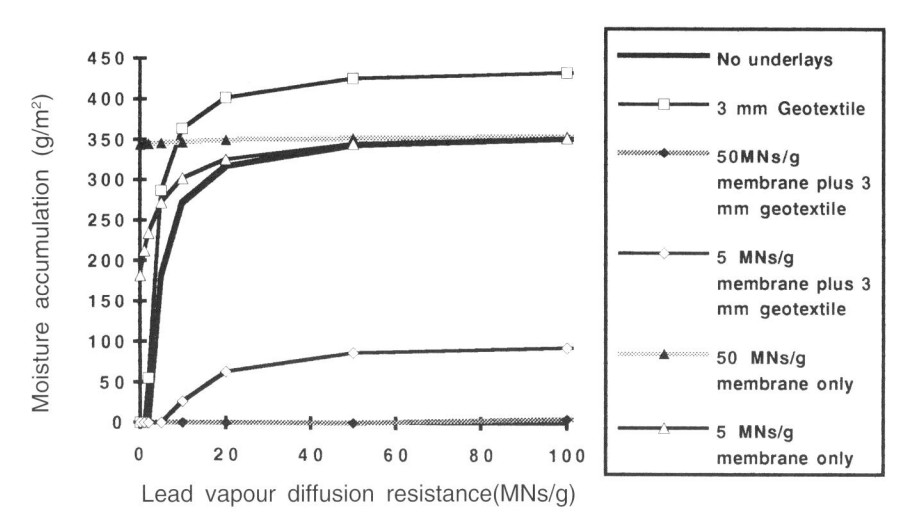

Figure 31. BS 5250 calculations for a variety of lead/underlay combinations. The six traces show the calculated accumulation of moisture in grams/square metre over a 60-day period averaging 15°C 65% RH inside and 5°C 95% RH outside, which is deemed characteristic of the build-up in a typical winter. The vapour resistance of the lead is varied to account for water vapour escape through rolls and laps.

Condensation risk

Condensation may occur on the surface of a material (as with dewdrops on the outside or the inside of a lead roof), or within constructions which remain superficially dry but inside which interstitial moisture accumulates.

Most conventional condensation calculations consider diffusion of water vapour in one dimension through homogeneous materials with constant levels of resistance to the passage of heat and water vapour, and under steady-state conditions of fixed internal and external temperatures, RHs and exposure levels. The BS 5250 calculation (*op cit*) works as follows:

- the types and thicknesses of the various layers of materials are determined.
- material properties are obtained from the tables in the BS, or from suitable references
- the thermal and water vapour diffusion resistance of each layer is calculated
- the temperatures and dewpoints at each interface between materials are determined
- if condensation planes are identified, then the rate of moisture diffusion to that plane is calculated by dividing the vapour pressure difference by the vapour resistance between the source of moisture and the condensing plane
- the moisture build-up in grams per square metre is assessed, over a 60-day period.

For lead roofs these calculations have many shortcomings:

- they consider moisture transfer by diffusion only, while air movement is important for lead, which is often laid upon discontinuous boarding, with permeable underlays and with opportunities for ventilation at the perimeter joints of each sheet
- they do not take account of dynamic wetting/drying effects with changing internal and external conditions, including sunshine and clear skies
- they do not take account of the ways in which properties of materials vary with moisture content, normally becoming more permeable the damper they are

Nevertheless, they are very helpful in giving an initial 'feel' for the important variables.

A series of of BS 5250 calculations have been undertaken using BRE's computer program BRECON II under the standard conditions (outside 5°C 95% RH, inside 15°C 65% RH) for Code 7 lead on 25 mm softwood boarding, assumed to be a homogeneous material with no air gaps between the boards. The underlay specifications are as follows:

1. no underlay
2. 3 mm geotextile
3. a reasonable vapour control membrane of water vapour diffusion resistance 50 MNs/g (MegaNewton seconds per gram), equivalent to about ten metres of still air
4. a poor vapour control membrane of water vapour diffusion resistance 5 MNs/g
5. as 3 with 3 mm geotextile between it and the lead
6. as 4 with 3 mm geotextile between it and the lead.

While lead itself is impermeable to the passage of water vapour, moisture will escape through the joints between the sheets. To illustrate the influence of this, the lead was given a range of water vapour diffusion resistance values from 0 to 1000 MNs/g.

Figure 31 shows the results of the calculations for a sheltered exposure. For all combinations, once the lead's diffusion resistance rises over 20 to 50 MNs/g (probably a reasonable average value in practice) there is little increase in the total amount of condensation over a 60-day period:

- the geotextile alone shows the most condensation because it offers little water vapour resistance but its insulating properties lower the temperature of the lead
- at high lead vapour resistances, the lead alone and the three membranes show similar amounts of condensation. At low lead resistances, condensation under the membranes is higher owing to their relatively lower permeability. Note that in these examples the condensation is under the membranes. While potentially this may protect the lead it could be dangerous for the membranes and for the underlying timbers
- the geotextiles over the membranes perform best, particularly with the more resistant membrane. How-

ever, this result has not been supported in site tests where ingressed and trapped moisture was troublesome in what amounts to a miniature 'warm' roof.

Figure 32 shows the effect of changing the indoor RH on moisture build-up and on BRECON II's calculated moisture content of the substrate boarding, again at 5°C outside and 15°C inside:

- at 15°C 50% RH the dewpoint is 4.8°C, and so there is no condensation
- as the RH rises, so does the timber moisture content and the amount of water accumulation below the lead
- the flattening-off at 85% arises because by this stage condensation is taking place not only below the lead but also under the timber decking: a similar situation occurs at a damp church in the Yorkshire Dales, after a cold, clear autumn night, and particularly under the colder tips of the copper fixing nails.

Although any moisture accumulation under the lead can be troublesome as far as corrosion is concerned, initially it will not appear as free water but will be absorbed by the wood. 25 mm thick substrate boarding with a specific gravity of 0.5 will weigh 12 kg/m^2. If its moisture content by weight were to increase from 10% to 20% from summer to winter, in principle it could absorb all the predicted condensate without entering the dangerous levels of over 20% moisture content. In fact, the situation is not this benign, because:

- moisture will be absorbed from the ambient air without condensation in any event
- moisture can get directly to the lead via the gaps between the boards. In completely still air, the amount transferred by diffusion is relatively small: water vapour only diffuses through still air about ten times as fast as through softwood; so if there are 3 mm wide gaps in 150 mm wide boards, the additional moisture accumulation under the lead would only be about 20%
- however, a much greater amount of moisture will normally get through by movement of moist air between the boards and out of the building via the rolls, accounting for the fish-tailing discussed in *Patterns of corrosion* above.

Dynamic modelling of moisture movement
The passage of moisture across the grain of wood can be a slow process; along the grain it can be a hundred times faster. Conversely, saturated wood takes a long time to dry out. In order to obtain a better idea of the importance of buffering by the substrate timbers, EH commissioned BRE Scottish Laboratory to undertake some test runs with the MATCH computer model. Like the BS 5250 method, this is also one-dimensional, diffusion-only. However, it models moisture movement through the timber more accurately by using different isotherms for adsorbtion and desorption and by having variable water vapour permeability. It also takes account of external climate changes: both air temperature and solar radiation on a time step basis, typically hour by hour.

Unfortunately, MATCH does not simulate indoor and roof space conditions. For the runs undertaken to date, a range of constant temperature and pressure rises above external ambient were assumed, typically 4°C and between 50 and 200 Pa (but restricted to a maximum roof space RH of 100%): above 200 Pa the lead was too damp for too long. At present the model does not take account of time lags between the interior and exterior environment or of the buffering and self-humidification and dehumidification processes which have been observed in roof spaces and discussed in *Roofs and roof space environments* above. It also does not simulate the drying of roof spaces by the sun in summer, although the effect of solar heating of the lead and the substrate boarding is included. It also does not appear to model accurately the effect of self-humidification when the sun shines upon the lead. Future modelling plans aims to improve these aspects by calibrating the model with roof space environment data from monitored sites.

The initial investigations looked at the variations in moisture movement for lead laid directly over oak, pine and plywood deckings, both for flat roofs and for pitched roofs inclined to the four points of the compass, and with different starting moisture levels in the boarding: 10% to represent dry timbers and 30% for wet timbers. The simulations ran for two years, starting either on 30 June (which from site tests appeared to be a good time to complete a lead roof) and 30 September (a bad time as it would rapidly encounter the autumn dews).

The most significant results were for the boards which started dry at 10% moisture content (M/C) on 30

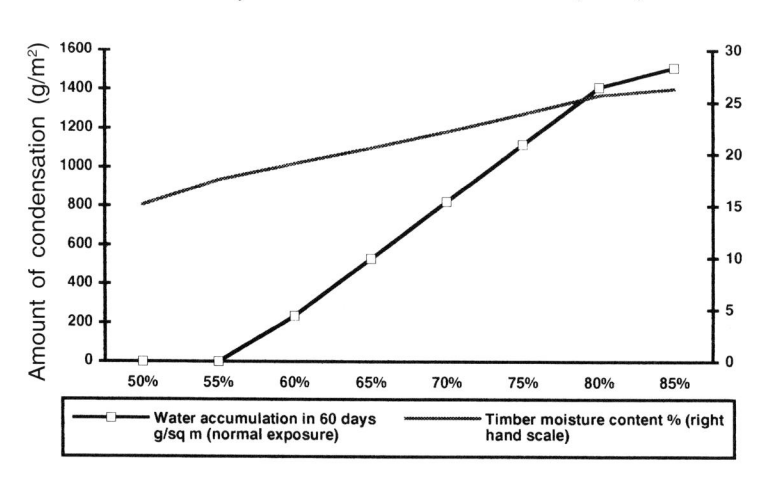

Figure 32. Amount of condensation over 60 days as a function of interior RH. At 50% RH the dewpoint of the indoor air is below 5°C and there is no condensation. The levelling off over 80% occurs because surface condensation also takes place under the boards: the calculated amount at 85% (1500 grams) is very similar to that taking place under the lead.

Plates

Plate 1. Abraham Darby's Iron Bridge at Coalbrookdale (1777–81) is a notable early example of the structural use of cast iron. From such precedents, cast iron warehouses, mills and factories were developed, many designed to be of fire proof construction.

Plate 2. Corrosion above softwood boarding. White and yellowish (States 1 and 2) corrosion product on the nave roof of a Buckinghamshire church, with dark passivation above the gaps between the boards. The photograph was taken after a cold, clear night, and the roof was particularly damp at the time. Note also the passivation over the rolls (with the odd spot of corrosion). The ridge is passivated on the far (right) side, probably by rainwater, which also seems to flow over the ridge from time to time (the lap is mean). Not surprisingly, there are signs of corrosion here, probably the result of a combination of condensation and refluxing of rainwater. In spite of this, and of the church being particularly damp (RTL 1993, Report 6, Appendix D), the corrosion is not yet serious. See also Figure 6, p 38.

Plate 3. Lead after a two-month autumn test at Donnington Castle showing a wide range of surface states from passivation to corrosion. The two (untreated) boards on the left have kept the lead dry and its underside is only slightly dulled. The two to their right were preservative-treated and probably initially damper. This has assisted passivation over much of the boards, but corrosion around the perimeter. See also Figure 7, p 39.

Plate 4. A church in Buckinghamshire: corrosion patterns on a slipped panel on the south aisle. See also Figure 8, p 39.

Plates

Plate 5. Lead over building paper on softwood on a relatively dry building. The lead is bright or only slightly dulled, though with a white haze between the boards. The pattern in the lap is not unusual: at the bottom a weathered area indicating capillary rise of rainwater, some corrosion (with a fringe of dark passivation) above that (where the rainwater distils) and the lead above that unaffected. The hollow rolls are free from corrosion. See also Figure 11, p 41.

Plate 6. Underside of lead over Erskine's felt at a mansion in Dorset. The lead over the underlying boards above the public rooms is lightly and fairly uniformly corroded. Above the gaps there is a tendency to passivation. This corrosion is no longer active. Note also the corrosion in the roll above the splashlap caused by distillation of trapped rainwater: this recurs on a freshly wire-brushed surface. See also Figure 12, p 42.

Plate 7. Underside of lead over Erskine's felt in a damper location at the Dorset mansion. The lead above the gaps is somewhat corroded, particularly towards the rolls in which there is also some corrosion. Note also the corrosion in the rolls above the splashlaps, as in Figure 12. The splashlap corrosion pattern varies with the distance apart of the sheets. Water is also trapped in the splashlap here by a slight backfall below the nosing. See also Figure 13, p 42.

Plate 18. Hard scales of corrosion. See also Figure 3, p 105.

Plate 19. Paint sample under 50x magnification. See also Figure 4, p 105.

Plate 20. Rapid paint deterioration as a result of continued underlying surface corrosion on unconserved railings. See also Figure 9, p 112.

Plate 5. Lead over building paper on softwood on a relatively dry building. The lead is bright or only slightly dulled, though with a white haze between the boards. The pattern in the lap is not unusual: at the bottom a weathered area indicating capillary rise of rainwater, some corrosion (with a fringe of dark passivation) above that (where the rainwater distils) and the lead above that unaffected. The hollow rolls are free from corrosion. See also Figure 11, p 41.

Plate 6. Underside of lead over Erskine's felt at a mansion in Dorset. The lead over the underlying boards above the public rooms is lightly and fairly uniformly corroded. Above the gaps there is a tendency to passivation. This corrosion is no longer active. Note also the corrosion in the roll above the splashlap caused by distillation of trapped rainwater: this recurs on a freshly wire-brushed surface. See also Figure 12, p 42.

Plate 7. Underside of lead over Erskine's felt in a damper location at the Dorset mansion. The lead above the gaps is somewhat corroded, particularly towards the rolls in which there is also some corrosion. Note also the corrosion in the rolls above the splashlaps, as in Figure 12. The splashlap corrosion pattern varies with the distance apart of the sheets. Water is also trapped in the splashlap here by a slight backfall below the nosing. See also Figure 13, p 42.

Plate 8. The original roof at the mansion in Buckinghamshire. Most of the roof was laid over Erskine's felt and was passivated between and corroded above the boards (the opposite of the current pattern). A few sheets, which did not have the felt, showed a similar pattern but with worse corrosion over the boards, probably because any moisture accumulating within this region would be trapped for longer than over the more permeable felt, and possibly there might have been some acidity. The longitudinal stripe was made by wire-brushing for tests and is not part of the original pattern. See also Figure 17, p 43.

Plate 9. Norwich Cathedral cloisters, showing lead distress and repairs. The pattern here is characteristic of acid attack, with not only weakening along the gaps between the boards, but also failure at the perimeter, at rolls and laps. See also Figure 18, p 44.

Plate 10. Underside corrosion over oak at Norwich Cathedral cloisters. The patch removed, on the left, failed at the perforation in the middle. The characteristic red/brown oxide can be seen near the interface with the metal, with the thick, flaky and granular corrosion product remaining underneath (on the right). See also Figure 19, p 44.

Plate 11. Chiswick House, London (English Heritage). See also Figure 1, p 83.

Plate 12. The decayed surface of the Bath stone gateway showing inappropriate render repairs (English Heritage). See also Figure 2, p 84.

Plate 13. The sphinx in its original location in the grounds of Chiswick House, but raised for lifting, with a stone copy in the background (English Heritage). See also Figure 1, p 95.

Plate 14. Underside of the sphinx showing the brick and mortar core with loose-fitted bronze stays (English Heritage). See also Figure 2, p 96.

Plate 15. Lighthouse-type measuring device (English Heritage). See also Figure 6, p 100.

Plate 16. Gateway and railings at Garrison Church. See also Figure 1, p 104.

Plate 17. Disruption of coping stone due to expansion of rusting stanchion. See also Figure 2, p 104.

English Heritage Research Transactions Volume 1

Plate 18. Hard scales of corrosion. See also Figure 3, p 105.

Plate 19. Paint sample under 50x magnification. See also Figure 4, p 105.

Plate 20. Rapid paint deterioration as a result of continued underlying surface corrosion on unconserved railings. See also Figure 9, p 112.

MATCH Simulation of Lead Roof
Initial moisture content of substrate = 10%
September 30th Start-up
Relative Humidity of outer layer of substrate

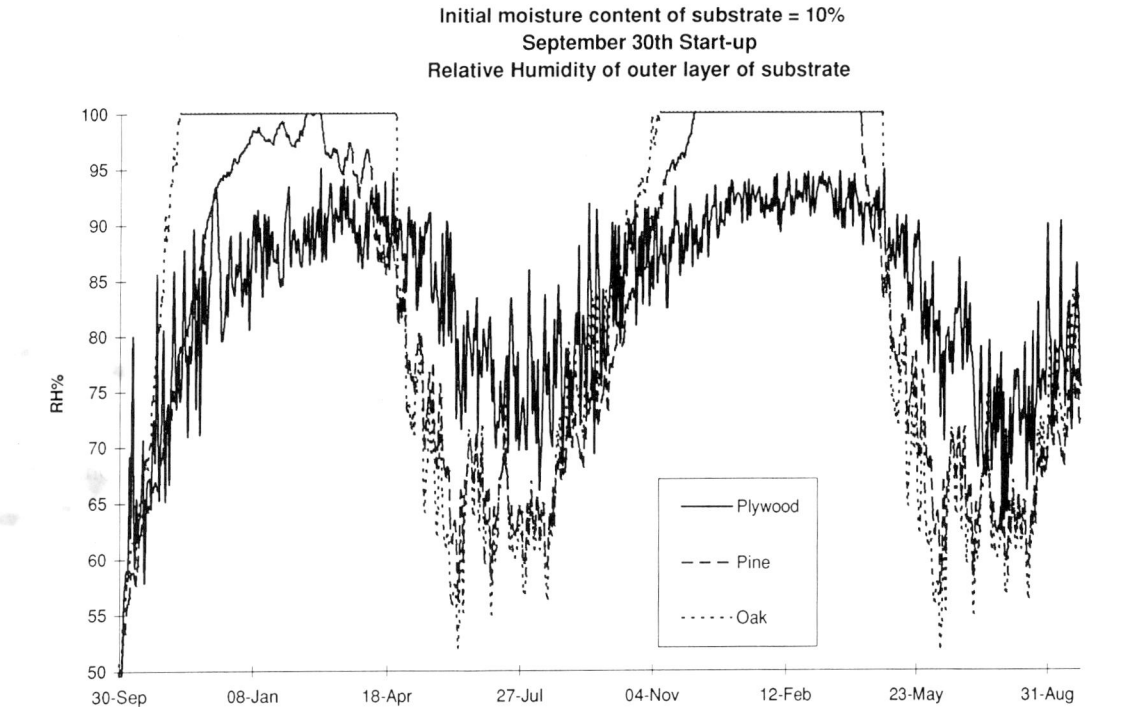

Figure 33. MATCH modelling of moisture content at lead/substrate interface.

September. Figure 33 shows plots of daily readings for a flat roof for two years, expressed in terms of RH at the outer layer of the substrate, immediately under the lead. The x-axis is the day number, starting on 30 September and running to 29 September two years later. With a 200 Pa/4°C vapour pressure/temperature gain, the lead/oak interface becomes saturated after some 40 days, while the pine gathers moisture more slowly with only a brief period of potential condensation after about 140 days, and the plywood never reaches saturation. Results for a north-facing roof were similar. For a south-facing roof starting at 10% M/C, for the roof completed on 30 September there was no condensation at all during the first winter and very little for one completed on 30 June.

In the second year, however, the pine never dries out to 10% over the summer and the interface becomes moist around Christmas and stays there until Easter. In practice, however, it seems that the model underestimates drying-out in summer: the 200 Pa vapour pressure gain is usually too high then because buildings are better-ventilated and drying in sunshine may not be adequately simulated. Certainly site measurements of the moisture contents of substrate boarding in September often show them to be in the region of 10%, and after a dry summer (as in 1995) sometimes considerably less.

The differences between pine and oak concur with findings on site, where in the first quarter of the year we have found visible water under lead over oak boarding and little or none over pine in the same building. Such differences in hygroscopic performance also help to explain the passivated stripe over the softwood fillet at Donnington Castle (see Figure 23). This is a consequence of oak having a higher water vapour permeability than pine when dry and a very much higher one when wet, so allowing moisture to accumulate faster in the hardboard above.

For plywood, however, it is not easy to relate the simulation data to site experience. Although we have found that over relatively dry buildings lead on plywood can perform well, in damper situations severe hygroscopic problems can occur, leading to the chemical problems discussed above. However, inspection of the source data showed that the MATCH database for plywood had only a 4:1 difference in permeability between wet and dry, while in the tests shown in Figure 25 the range was ten times greater! The plywood was also simulated as a homogeneous material: different results would be obtained by modelling it as porous timber separated by less pervious glue lines. In addition, a lot depends on the source and batch of the plywood. Further studies are recommended.

In a second batch of studies, BRE looked at the number and duration of wetting/drying events for an oak roof, starting with a 30% moisture content on 30 June, for vapour pressure differences (overpressures) between 50 and 200 Pa. The statistics shown in Table 3 are interesting.

Essentially, for overpressures between 50 and 150 Pa the number of annual condensation events is similar at about 30: what changes is their duration. Little significance should be read into the precise numbers, which fluctuate rapidly with small changes in overpressure, as events separate or merge. At 50 Pa there is a rapid succession of very short events, typically in late December. Above this overpressure the average duration ceases

Table 3. Results from dynamic modelling.

OVERPRESSURE (pascals)	50	100	150	200
Percentage of year condensation is present	1.4	14.3	26.5	47.3
Number of events per year	32	37	25	10
Average hourly duration of wetness period	4	34	126	415

to be very meaningful owing to a long period of winter wetness with most of the events taking place in the autumn (as the roof moves into wetness) and in the spring (when it moves out again). For example at 100 Pa, there are six cycles in late November, as the roof begins to get wet, none during December or early January, a rapid succession of 25 cycles over a fortnight in late January and the final few in late February. At 200 Pa the events associated with wetting occur in late October, with drying-out events not until April.

Although only preliminary work was done, and with some reservations about the results, the MATCH modelling does support the evidence from sites that differences in the material properties of the substrate timbers can have major effects on wetting, drying and corrosion/passivation processes immediately under the lead. In addition, in particular for the pine, the time of laying, the initial moisture content and summer drying-out processes can significantly influence the dampness under the lead in the autumn, and perhaps for the whole of the subsequent winter. Where pine starts dry or has been dried out in the summer, a slow transition from dry to wet may also allow passivation to occur during the intervening moist, non-condensing conditions. More detailed investigation is about to start, including some full-scale laboratory tests.

The wetting/drying events analysis is also of interest. RTL's work suggests that in a neutral environment with little organic acid, passivated lead surfaces can withstand some 50 or more cycles events before they begin to break down. If the preliminary MATCH results are substantiated, then there may seldom be this many cycles in a year. If therefore:

- the lead is passivated initially
- the substrate material has suitable hygroscopic and chemical properties
- the substrate material starts out dry
- the environment allows passive films to repair themselves when conditions change

the prospect of lead being able to resist some wintertime condensation is offered. This supports the evidence from a number of sites.

Conclusions

This section has only scratched the surface of the problem but offers some interesting avenues for future investigations and analysis. It confirms the importance of the substrate material and its initial condition and suggests possible methods of minimizing the underside corrosion risk to lead in situations where condensation may occur from time to time. However, one must remember to protect not only the lead but the building as a whole. At present it seems unlikely that the right combination of substrates, underlays, chemical treatments and conditions at the time of laying would be able to provide suitable protection from winter-time over-pressures much in excess of 150–200 Pa. Beyond this, explicit attention to heating, ventilation, moisture removal or roof construction may well be necessary.

Promising future work would include:

- further modelling of hygrothermal properties of the lead/underlay/substrate system
- detailed consideration of the different relationships between indoor and roof space environments, for the four different roof types
- modelling of hygrothermal buffering in roof spaces and the benefits or otherwise of outside air ventilation
- more timber moisture measurements, if possible including continuous recording of seasonal changes
- review of the adsorption/desorption, water vapour transmission and acidity characteristics of a variety of timber products, both theoretically and in the laboratory, to identify which are most appropriate as substrates
- analysis and interpretation of RTL's monitored temperature/RH data, in particular to understand typical overpressures
- theoretical and laboratory review of the beneficial and adverse effects of different boarding types, gap sizes and underlays on condensation, corrosion and passivation
- further review of suitable underlays.

6 Discussion

The sections above have outlined some aspects of the chemistry of lead, roof construction, heat, air and moisture movement in roof spaces, and site evidence of underside corrosion. All of them reveal a capricious system, with a high variability which often depends upon small differences between larger, and often unknown quantities.

Underside corrosion in context

Underside corrosion is nothing new. It has been known for centuries to be a contributor to the decay processes which cause lead roofs to need replacing from time to time. Although some lead roofs last 80–200 years, and occasionally even more, records for some churches visited suggest that on occasions only 40 years or so passed between major re-leadings: so some problems they have now may well have occurred in the past too.[30] Since we are examining them more now, we are also more alert to early underside corrosion (some of which is only cosmetic) and to incipient failures which would previously have gone un-noticed.

Lead is durable, but not infinitely so, a feature it shares with nearly everything else. But in these days where many things claim to be maintenance-free (although in practice this may mean 'impossible to maintain', we may be expecting more of lead than we did. While lead lasts longer than many alternatives, its properties need to be respected, just as glass, a much more fragile material, can give a long and largely maintenance-free life if properly specified, handled, installed and used.

While condensed water is enough to cause underside corrosion, where the lead has actually failed rapidly, aggressive chemicals have nearly always been involved. Many timbers, and the vapours from them, can harm lead, particularly if they get wet. Today this may be exacerbated by preservative treatments (if only through increasing the timber's initial moisture content), by kiln-drying, and by a wider range of timber sources. In historic buildings the most common offender is oak, whose aggressive effects have been known for millennia. Nevertheless, this sometimes seems to be tacitly ignored, occasionally for new oak and more often for old, on the assumption that it will have lost its aggressiveness. This is not necessarily so, particularly where conditions have become moister. The studies to date also indicate that barrier layers intended to separate lead from such timbers are seldom totally effective, and that the potential aggressiveness of modern wood-based sheet materials is not always realised.

Is underside lead corrosion getting worse?
Over the past 50 years or so, changes in the heating, ventilation, occupancy and insulation of buildings have tended to make roof spaces damper, and the 1939–1945 war no doubt helped through shortages and neglected repair and maintenance. However, the study has shown that the relationship between these environmental changes and increased corrosion is not as direct, or as reversible, as was thought.[31] While there is good reason to think that the number of failures, and not just the awareness of them, did increase during the 1970s and early 1980s, these seem to have been as much due to new substrate materials and details as to altered environments, though increased dampness has certainly not helped. However, actions already taken have brought some of the problems under more control.

Ten years ago it became known that the 'warm' roof principles widely applied to roofs with continuous membranes did not suit continuously-supported metal roofs. The main reason is that moisture entering from any source becomes trapped. This trapped and refluxing moisture, which also tends to pick up organic acids and other chemicals, can affect many metals but is particularly aggressive to lead. It is now widely known that this form of roof construction should not be used for lead, and so this source of severe underside lead corrosion will be progressively eliminated.

While the physics of thermal pumping can affect any roof, the use of splashlap details on 'warm' lead roofs, and particularly low-pitched ones, made them particularly susceptible to sucking in rainwater. Rainwater can both cool the roof rapidly, causing the trapped air to contract, and also seal the splashlaps by capillary attraction, closing air paths via the joints which would otherwise have provided pressure relief. The ventilated warm roof now recommended avoids this problem.

Another cause of early underside corrosion failure was the use of has been deckings of wood-based panel products such as plywood. Their raw materials are often more acid than the traditional softwood and resins, glues and processing can make them more acid still. While they can perform well in dry environments, in damper ones failures may be much more rapid. Their physical properties may also increase the incidence of moist, corrosive conditions at the lead/substrate interface. For a while these materials were also used in composites under very thin lead, and these were subject to corrosion, thermal fatigue and combinations of the two. Although these composites are no longer sold, and in historic buildings there has been a trend back to softwood boarding, manufactured boards may still sometimes be used in inappropriate circumstances.

By avoiding 'warm' roof construction, and using ventilated warm roofs where appropriate, the circumstances that led to the most severe corrosion failures (apart from those in very humid and/or aggressive environments such as swimming pools and some industrial processes) are being eliminated. However, the classic problems in historic buildings still remain, sometimes exacerbated by damper environments under the roof. The research has also found that ventilated warm roofs and well-ventilated roof spaces are not always immune from some underside corrosion.

Improving the situation
Laboratory, field and theoretical studies have all shown that the state of the lead's surface when it first encounters moisture can have a major influence on its long-term corrosion behaviour, particularly in marginal circumstances. Contact of fresh, clean lead with rain, condensation or a damp substrate must be avoided and the early formation of a protective film should be encouraged. In the past, some protection may have been provided by one or more of:

- leaving the lead lying around for some time (not tightly-rolled) before fixing. However, this gives only limited protection. Slight protection may also be offered to the rough side of cast lead by being steamed on the sand bed
- applying linseed oil to the lead either deliberately (by wiping) or incidentally (as a lubricant in the rolling-mill)
- laying the lead on a relatively dry substrate at an appropriate time of the year, preferably between May and July. In the past plumbers would have tried to re-roof when the weather was better, now with project approval often tied to financial years, contractors report that this can be their slackest time for lead roofing work.

In the laboratory the simplest and most durable means of pre-passivation found to date has been to paint the lead with a slurry of chalk powder in water: and this may allow higher initial timber moisture levels to be tolerated. Site tests must continue, which look not only at simple flat specimens but also at the wide range of geometries and details that can occur.

Ventilation is often proposed as the cure-all, but its function and purpose is often misunderstood and consequently mis-applied. There are three main ways in which ventilation may be beneficial:

- ventilation of the building as a whole helps to remove moisture by displacing inside air by outside air, which on average will tend to have a lower dewpoint. However, the same air also removes heat in the colder weather when condensation and dampness is more likely. Unless this heat is put back, the temperature will drop and the relative humidity will increase, undoing much of the benefit. Indeed, if the source of moisture remains but with little or no extra heat, in spite of a reduced dewpoint the building may actually become damper, as has been found in many attempts over the past twenty years to improve conditions in low-income housing by ventilation alone. Alternatively, if the temperature is maintained and the ventilation is increased too much, then in cold weather the intense atmosphere may become too dry, which although perhaps good for the roof may well cause problems with shrinkage and cracking of organic materials such as timber and efflorescence and crystallization damage to masonry. In general, occupiers will not pay to heat what they regard as excessive amounts of ventilation and so a compromise will often fall short of what might be seen as best for the roof. While an adequate level of heating and ventilation should be aimed for, this level will seldom be sufficient to keep the roof dry enough to be out of danger. The traditional combination of high fuel consumption, high ventilation rates and relatively low air temperatures obtained with many open fires is no longer sustainable.
- ventilation of the roof space under the lead and its decking. The ideal of the 'cold' roof, in which there is so much ventilation by outside air that all moisture gains from within the building are borne away with no significant increase in dewpoint, is most closely approached in generously-ventilated roofs such as bell-towers (provided that rain does not blow in and air leakage from the church through the floor is very much less than the cross-ventilation through the louvres), and with less but well-distributed ventilation in well-detailed ventilated warm roofs with effective air and vapour control layers. The moisture content of the timber will depend on the climate, the time of year, the hygroscopicity of the timber species, its size and shape, the heating and ventilation of the roof space, and so on, and with variations for the individual sample. Owing to its lower temperature, 'air-dry' timber here will be relatively damp in winter (see Figure 5), so any small amounts of additional moisture may be sufficient to wet the underside of the lead. At 90% relative humidity, a hygroscopic species such as Corsican Pine typically has an equilibrium moisture content of 22% by weight [32], European Redwood (and European Oak) 20% and Yellow Pine 15% (Stillman and Eastwick-Field 1966). RTL suggests that underside corrosion can start to occur at timber moisture contents over 20%, so it may be worth evaluating timbers of different hygroscopicity and vapour permeability
- Ventilation to the underside of the lead itself. This often occurs through the gaps (penny or larger) between the underlying boarding. It can be a mixed blessing, as we have seen in *Patterns of corrosion* above: sometimes the lead here is corroded, sometimes passivated. From the ventilated warm roofs studied and the 'Dutch Barn' external test rig it seems that in a fully-ventilated situation as outlined above, the underside of lead exposed to the open air often falls slightly below the dewpoint and only just avoids corrosion in ambient conditions: small increases in vapour pressure (as with an imperfect vapour control layer) or small decreases in temperature (as by evaporative cooling from a puddle in a slight depression in a dead flat section of roof), and possibly even a somewhat colder or damper climate, are sufficient to initiate underside corrosion. Unless the roof space ventilation corresponds to near-ideal conditions, its effectiveness in protecting the lead must be questioned, and alternative or supplementary measures may be required. RTL find that a layer of plain building paper under the lead can give useful physical protection against transient diurnal condensation, without trapping ingressed moisture. Chalk treatment may be able to provide additional chemical protection. Tests are continuing.

Laying the lead on a gapped substrate has one very important advantage: even though condensation corrosion may occur in adverse conditions, the worst situations associated with trapped, refluxing and acid-containing moisture may be avoided. If chalk treatments prove to be successful, they can create more opportunities for substrates which restrict the access of air and moisture to the underside of the lead. Further studies are recommended, both theoretical and practical, of material properties, moisture movement and whether underside corrosion can be prevented both on site and in the laboratory.

While chalk coatings have been promising, they need further testing and development for reliable use on site. In particular this includes:

- Testing of suitable, low-cost commercial supplies of powdered chalk in the laboratory and on site
- developing reliable guidelines and procedures for coating the lead on site
- consideration of additives which might reduce the risk of attack by carbonyl compounds etc. Chalk itself provides some initial protection from acetic acid, but cannot resist high concentrations, as from fresh oak, and its long-term performance is not known
- developing and testing suitable underlay specifications (see below)

- considering and developing pre-loaded underlays. Site tests indicate that these would be additional to, and not instead of, the chalk slurry coating
- reviewing how well lead can be protected where it is not in direct contact with the substrate boarding, especially where it turns into the rolls where fish-tailing can occur over gapped boarding and acid/distillation-related corrosion in more sealed environments
- longer-term tests of durability and self-repair in a chalk-rich environment.

CAUTION: chalk treatments look promising, but are not yet commercially available for use by specifiers because their performance has not yet been proven at full scale. Full-scale tests are being undertaken in 1996-97 on some English Heritage and Historic Royal Palaces Agency sites where areas of roofing are being replaced or repaired. Pre-treated underlays (the concept is already subject to a patent by English Heritage) are being promoted by some suppliers, but whatever they may say these have not been developed, approved or tested by or in association with English Heritage and have not been endorsed by the research.

To date underlay investigations have been disappointing: in moist situations permeable ones let water vapour in, impermeable ones trap any moisture going, which is worse, and double-layer underlays have been dissappointing. More work is required on suitable underlays to carry the chalk layers, including possible composites, in which the upper part is sufficiently open to carry the chalk while the lower part both stops it falling through and controls to some extent the ingress of too much water vapour and moist air.

The literature review and theoretical analysis has shown that outside-air ventilation of roofs to which sources of moist air and water vapour from the building underneath have not been stopped (or at least very severely restricted) can be of little use, or worse, in avoiding condensation. Indeed, where gains are not too large, roof spaces with limited ventilation may sometimes be useful reservoirs in which moisture accumulates only slowly. In the summer, even where ventilation is restricted the roof may dry out effectively while in the process the buffered environments may help to passivate the lead (or repair partly-eroded passive films), and protect it from transient condensation. It is not yet clear how these mechanisms can be used reliably to positive effect. Further investigation and analysis should continue.

Much data has been collected in the buildings already visited. While partially interpreted, it could yield more information. To avoid too much dispersal of effort, it would be better to continue analysis and testing on some of these sites than to introduce new ones.

Giving preliminary guidance to specifiers
A separate document is being prepared to update architects and surveyors. It will:

- outline the findings of the research to date
- re-state the principles of the ventilated warm roof and draw attention to the need for meticulous attention to detail in air and vapour sealing and to providing effective through-ventilation
- caution against the use of inappropriately damp or acid substrate decking
- identify an approach to diagnosis of underside lead corrosion problems and testing possible solutions.

While there is clearly a strong need and demand to move some of the findings from the research into full-scale pilot application, to date this has proved troublesome owing to:

- the absence of definitive guidelines, other than for ventilated warm roofs
- excessive enthusiasm by some in the industry about the use of chalk coatings and underlays
- issues of professional indemnity in making use of ideas which inevitably have not yet been thoroughly tested in practice.

A standard approach to such cases needs to be developed, including:

- acceptance of some of the technical risks by the client
- a clear statement that any unusual specifications are for testing and development and not definitive new practice. A standard letter to this effect has now been agreed by English Heritage, the Historic Royal Palaces Agency and the Lead Sheet Association.

APPENDICES

A: Carrying out sample tests

If a roof has not been subject to significant underside lead corrosion, past advice has been that, provided nothing else changes, it would be reasonable to re-cover it in a similar manner, perhaps with some additional ventilation. However, the research has shown that even direct replacements do not always perform in the same way. There are several possible reasons for this:

- the environment now may be more aggressive than it was, but the existing lead has become passivated during its life, and this is now protecting it
- the new lead may be laid in unfortunate conditions (for example in the autumn, or on a wet building), which predisposes it to corrosion
- small changes (eg in substrates, underlays, detailing or ventilation) may be critical.

The research suggests that, for freshly-prepared samples, the type of lead has very little influence on the initially-observed corrosion.[33] Some tests are therefore possible merely by wire-brushing patches of the underside of the existing lead, taking appropriate safety precautions against ingestion or inhalation of the resultant dust (Lead Sheet Association 1993c), to expose a clean surface. The sample areas should include as wide as possible a range of conditions, and in particular:

- any locations showing systematic evidence of some underside lead corrosion (but also locations which do not, these may not necessarily be inert now)
- in the centre of a bay and at the edge
- over gaps and other weaknesses in the underlying decking (if present).

Photographs of typical underside corrosion patterns and sites can be found in *Patterns of corrosion* above.

If new lead, or lead with alternative pre-treatments, is to be tested, a simple procedure is merely to lift the existing lead, lay the samples, typically about 100 mm square, underneath in positions as outlined above, and re-lay the lead as a capping sheet. Ideally, thermally-conducting paste would be placed on the top of the sample to give better heat conduction to the outside, but in the research it was found not to be necessary. Alternatively, a whole bay or bays can be replaced: this may be a sensible option where a sheet has failed anyway and needs urgent replacement, or where the substrate is also to be renewed.

For either existing or new lead samples, different substrates and underlays can be placed underneath as required. However:

- larger samples will normally be necessary to permit all conditions to be monitored
- if underlays are intended to be vapour-resistant, steps must be taken to avoid water vapour getting around the sides.

As a general rule, lower-pitched roofs are more prone to corrosion than steep ones, very well-ventilated roofs (such as in bell towers) are less prone than elsewhere, roofs with roof spaces (ventilated or not) are less prone than those without, and for those without, roofs at high level are more prone than those at low level. Sometimes there are also significant variations with position and orientation, see *Patterns of corrosion* above.

Ideally samples should be first set up in May. An inspection in late September will then reveal whether any corrosion or passivation has occurred in the summer. At this time, half the area of each sample (or one sample where pairs are being used) should be wire-brushed (and this area re-coated, for example with chalk slurry, if this was done in May). The autumn is often the worst time for underside corrosion and the samples should be checked again around Christmas: sometimes the September ones will be corroded and the May ones not. Finally, at around Easter the two parts of each sample can be compared. If no samples of a particular type are corroded (one hopes that chalk treatment may often do this), then one can proceed with caution, though care will still need to be taken to keep the site dry during laying. Otherwise, more thought will be required.

B: Members of the Condensation Corrosion Forum

The Condensation Corrosion Forum has met annually during the project to review the conclusions and to coordinate them with other research and expertise.

- Professor Geoffrey Allen, University of Bristol, Interface Analysis Centre: conducting research into fundamental processes in the corrosion of lead (joint project with SERC and The National Trust)
- Dr Paul Baker, Building Research Establishment, Scottish Laboratory: conducting research into moisture movement and condensation in buildings and roof constructions
- Leo Biek, chemist: consultant and former head of the Ancient Monuments Laboratory at English Heritage
- Leon Black, University of Bristol, Interface Analysis Centre: conducting research into fundamental processes in the corrosion of lead (joint project with SERC and The National Trust)
- Stephen Bond, Historic Royal Palaces Agency: involved in the maintenance, repair and replacement of a large number of lead roofs, co-client for RTL's research
- Dr Bill Bordass, William Bordass Associates: chairman and coordinator of English Heritage programme
- Dr James Charles, University of Cambridge, Department of Materials Science: conducting research into the differences between milled and DM cast lead
- Dr Rob Edwards, Department of Chemistry, Liverpool John Moores University: conducting research into corrosion of archaeological lead and providing analytical support to the English Heritage project
- Dr David Farrell, Rowan Technologies Ltd, Manchester: main contractor for English Heritage site and laboratory tests
- David Farrington, Historic Royal Palaces Agency: involved in the maintenance, repair and replacement of a large number of lead roofs, co-client for RTL's research
- Paul Frost, Calders Industrial Metals Ltd: expert on the performance of lead, providing scientific and analytical support
- Neil Lewis, Lead Sheet Association and Calders Industrial Metals Ltd: lead industry representative and technical liaison with LSA
- Chris Sanders, Building Research Establishment, Scottish Laboratory: conducting research into moisture movement and condensation in buildings and roof constructions
- Dr Nigel Seeley, The National Trust: chief scientist
- Chris Wood: client for English Heritage's work
- John Woods, Lead Sheet Association representative

Contributors to earlier meetings of the Forum:

- Iain McCaig, initial client for English Heritage
- Ray Cox, Building Research Establishment, Metals Section: expert in the performance of metals in buildings
- Brian Day, University of Bristol, Environmental Engineering Studies Unit: conducting research into moisture movement and storage in building materials
- Philip Forshaw, University of Bristol, Interface Analysis Centre, Research student on lead corrosion
- Professor Jack Harris, University of Bristol, Interface Analysis Centre

- Dirk Janssen, Rheinzink: providing information on continental European practice for continuously-supported metal roofs

C: Research projects

- Laboratory tests of corrosion under intermittent wet/dry cycles: Rowan Technologies Ltd
- Outdoor full-scale test rig: Rowan Technologies Ltd
- Site investigations, sample tests, and overall reporting: Rowan Technologies Ltd. Computer modelling, site studies and test rigs at BRE Scottish Laboratory

Support and advice received from Dr Bill Bordass (William Bordass Associates), The Lead Sheet Association, The Historic Royal Palaces Agency, The National Trust, Borough of Preston, SAS Software Ltd, Bickerdike Allen Partners, Follansbee Ltd, Ove Arup Partnership and Vis Williams Partnership.

D: Sites visited (see table 4 for key characteristics)

- *Church 1 in Yorkshire: observation and tests*
  The nave has oak boarding which is both the ceiling and supports the lead. It was re-roofed in 1938 with a separating layer of bitumen-cored building paper, which appears also to have been bedded in bitumen over the oak. The Code 8 sand-cast lead in some areas, especially on the apexes and at the edges of the rolls (a typical weak spot where acid-related corrosion is involved), has now corroded through. In these positions the building paper had also tended to fail.
  Conclusions: An example of the difficulty of protecting lead from the effects of acetic acid in the long term. Chalk coatings, which have been found to give some protection in the laboratory, are now being tested here.

- *Church in Buckinghamshire: observation, monitoring and testing*
  Lead on the roofs to the tower and the two aisles date from the nineteenth century, the nave was re-leaded in 1939. The sheets on the aisles have slipped over the years and are now admitting water in places. All roofs are laid on softwood boarding with gaps averaging some 5 mm. As in many village churches the underside of the boarding forms the ceiling of the church. The underside of the lead on the well-ventilated tower roof (now also used as the air intake chamber for the heating) is in good condition. The other roofs showed some underside corrosion, related to the gaps between the boards but which had varied over the life of the roof, in places passivating and sometimes corrosive. Monitoring has shown that the atmosphere in the church is currently very damp, and condensation events frequent, even in summer. Sometimes condensation drips onto the pews, particularly in the nave: probably owing to the stratification of moist air. Given the amount of dampness, it is surprising that underside corrosion was not more severe.

  Pressurised direct gas-fired heating was installed in the late 1980s. It is likely (though not entirely certain) that combustion moisture from this may have exacerbated the condensation and corrosion: some corrosion, particularly to the nave roof, appears to have started within the past few years. Tests and environmental monitoring are continuing.
  Conclusions: Confirms that there is no direct relationship between the amount of moisture and the amount of corrosion. As a general rule, flueless heating should not be used where there are lead roofs. The south aisle roof is currently being used both as a test site and for environmental monitoring: more information will be available from RTL in due course.

- *Brightling Observatory, Sussex: observation*
  Domestic background heating. A house on an exposed hilltop which has slate roofs with lead flat tops, gutters and window cills. 'Cold' uninsulated roof spaces quite well isolated from rooms underneath by thick lath and plaster ceilings with little cracking and few holes for services, hatches etc. No explicit ventilation but adventitious ventilation via slates, particularly when windy. Underside of lead directly on softwood close boarding in good condition, even over rotted timber in gutter. Some corrosion under window cills and flashings where lead had been stuck down with dabs of acrylic (?) sealant after it had been lifted in gales (1987?).
  Conclusions: Probably no need for explicit ventilation when the slate roof is repaired, provided no sarking felt is added under the slates, and if patination oil or chalk pretreatment is used. A separate roof on the tower should be of ventilated warm roof construction.

- *Caerhays Castle, Cornwall: observation, monitoring and tests*
  Domestic background heating. The lead dates from the 1850s and is now in poor condition from thermal fatigue. Underside corrosion present on some sheets but relatively thin in spite of damp roof space, timbers and external environment. In a few places, the softwood boarding had been replaced by elm when repaired in the 1970s (?): lead over some of these boards (typically with acetate content of 80 ppm or more) has corroded. The mild, moist microclimate could well be passivating at times. The roofs get extremely hot in the sunshine, owing to shelter from the hillside and trees behind and absorption by a dark brown topside patina (probably an effect of the microclimate and possibly sea salt).
  Tests showed that:

  - lead wire-brushed in September corroded over gaps and holes but over one cracked sheet, which let water in, the wire-brushed lead had passivated, even some distance away where there was visible condensation from the refluxed moisture
  - lead wire-brushed in September and treated with chalk paste did not corrode, over softwood, the

Table 4. Lead roofs inspected in the UK.

NAME	Variant	Location	Building Type	First built	Lead dates from	Building usage	Heating type	Heating schedule	TEMPERATURE when heat on. Scale 5=warm to 1=chilly	TEMPERATURE at other times. Scale 5=warm to 1=chilly	VENTILATION: Scale 5=liberal to 1=limited	DAMPNESS Scale 1=dry to 5=very damp	Classification	Ceiling	Extra vapour control layer	Added insulation	Void	To room air	To outside air	Likely main origin of roofspace ventilation
Brightling Observatory	Typical	Sussex	House	1810	1900s	Domestic	HW rads	Domestic	4	2	2	4	3 Domestic	Plaster	None	None	Crawl	Adventitious	Via slates	Outside?
Caerhays Castle	Typical	Cornwall	Mansion	1808	1850s	Domestic	Night storage	Background	3	3	3	3	4 U'drawn	Plaster	None	None	Shallow	Adventitious	Via loose lead	Indoors
Cathedral in North West	Existing Elm boards	Northwest	Cathedral	>500 yrs	1980s	Cathedral	Radiators?	Daily	4	3?	3	3	2 ??	Wood?	No	No	No	Not known	Adventitious	Indoors
	New VWR												VWR	Wood?	Hyload	50 mm	50 mm	Vapour sealed	Top and bottom	Outside
Church in Bucks	Aisle	Bucks	V church	>500 yrs	1930	V church	Direct gas	Occasional	4	2		3	5 Direct	None	None	None	None	Complete	Outlet via gaps only	Inside
	Nave				1939								Direct	None	None	None	None	Complete	Outlet via gaps only	Inside
Church in Northants	Nave	Northants	V church	>200 yrs	1988	V church	HW radiators	Continuous	4	4	4	4	1 Direct	Timber	None	None	Shallow	Complete	Adventitious	Inside
Church near Sheffield	Existing	S Yorks	Church	>200 yrs	Varies	V church	Floor trench	Continuous	4	4	1	1	5 U'drawn	None	None	None	None	Complete	Outlet via gaps only	Inside
	New VWR				1995		HW radiators						VWR	Wood	Yes	Yes	50 mm	Vapour check	Top & bottom	Outside
Church in Shropshire	Aisle	Shrops	V church	>500 yrs	1962	V church	HW pipes & rads	Daily	3	3	3	2	2 U'drawn	Plaster	Bit felt	none	Shallow	Adventitious	Adventitious	Minimal
Church 1 in Yorkshire	Nave	S Yorks	V church	>500 yrs	1938	V church	HW fan conv'rs	Occasional	3	2	2?	2?	3? Direct	None	Bitumen felt	None	None	Complete	Outlet via gaps only	Minimal
Church 2 in Yorkshire	All	North Yor	V church	>500 yrs	1990	V church	HW pew rads	Occasional	3	1		3	5 Direct	None	None	None	None	Complete	Minimal	Inside
Church 3 in Yorkshire	Chancel	S Yorks	V church	>200 yrs	?	V church	HW radiators	Occasional	3	1		3?	4 Direct	None	None	None	None	Complete	Outlet via gaps only	Inside
	Aisle												Direct	None	None	None	None	Complete	Outlet via gaps only	Inside
Civic building	Typical	Herts	Amenity	1980	1991	High	HW rads	Daily	5	4		3	3 VWR	Timber	Polythene	50 mm	15-20 mm	Vapour check	Top and bottom	Outside
Donnington Castle	Existing	Berks	Monument	>500 yrs	1955	None	None	None	5	4		3	3 VWR	Timber	Polythene	50 mm	15-20 mm	Vapour check	Top and bottom	Both
Educational building	Typical	Cambs	Common Rm	1966	1966	Meeting	HW convectors	Daily	1	1	4	4	4 Direct	NA	NA	NA	NA	Complete	Outlet via gaps only	Inside
	Facia								4	3		4?	2 U'drawn	Timber	Slaters felt	25 mm	Shallow	Via cracks	Adventitious	Both
													Facia	Timber	Slaters felt	25 mm	Shallow	Via cracks	Adventitious	Outside
Hampton Court	Gt Hall	London	Great Hall	500 yrs	1955	Visitors	Underfloor	Continuous	3	3		3	2 U'drawn	Wood	None	No	Shallow	Via timber joints	Adventitious	Inside
Manchester Cathedral	Aisle	Mchester	Cathedral	>500 yrs	1964	Cathedral	HW radiators	Daily	3	3		3	2 U'drawn	Wood	None	none	Shallow	Loose fit ceiling	Adventitious	Inside
Mansion in Bucks	Original	Bucks	Mansion	1906	1906	Office	HW radiators	Daily	5	4		2	1 Domestic	Plaster	None	50 mm	Crawlable	Adventitious	Adventitious	Inside
	New				1988	Office			5	4		2	1 Domestic	Plaster	None	50 mm	Crawlable	Adventitious	Adventitious	Inside
	Separate			1906		Stables	Air conditioned	Daily	4	4		5	1 U'drawn	Cities	Polythene	50 mm?	Shallow	Poor vap check	Adventitious	Inside
Mansion in Derbyshire	Existing	Derby	Mansion	>200 yrs	1900?	1986 Training rm Museum	HW radiators	Constant	4	4		2?	2 Domestic	Plaster	None	None	Walkable	Adventitious	Adventitious	Inside?
	Renewed				1992	N Trust hse	Dehumidified	Conserv'n	2	2		3	4 Cold	Plaster	Yes,poor	Slabs	Walkable	Adventitious	Eaves and apex	Both
Mansion in Dorset	West end	Dorset	Mansion	>200 yrs	1984	N Trust	HW radiators	Domestic	5	4		3	2 Cold	Plaster	None	None	Walkable	Adventitious	Hatches+tubes	Inside
	East end							Conserv'n	3	3		3	2 Cold	Plaster	None	None	Walkable	Adventitious	Hatches+tubes	Inside
	Mansard							Domestic	5	4		4	2 Cold	Plaster	None	None	Varies	Adventitious	Tubes	Mixed
Mansion in Northants	Renewed	Northants	Mansion	200 yrs	1994	Medium	HW radiators	Domestic	?	?		2?	2 Cold	Plaster	Foil back?	100 mm?	Shallow	Adventitious	Via new turrets	Minimal
	Chapel								3	3		2	3 U'drawn	Plaster	Bit felt	none	Shallow	Adventitious	Adventitious	Minimal
Metal store		London	Castle	>500yrs	1930s	Metal store	Radiator?	Constant	4	4		3	2 U'drawn	Plaster?	None	None	Shallow	No explicit	No explicit	Inside
Museum		London	Castle	>500yrs	1967	Museum	HW fan conv'rs	Constant	5	5		4	2 U'drawn	Timber	None	None	Crawlable	Exit route	Fan louvres	Inside
Norwich Cathedral	Choir sch	Norwich	Cloister rms	>500 yrs	1953	Daily	HW radiators	Daily	3	3		3	2 Direct	None	None	None	None	Complete	Outlet via gaps only	Inside
	Refectory		Refectory	>500 yrs	1956	Daily							U'drawn	Plasterbd	Foil back	25 mm?	Shallow	Adventitious	Swan necks	Not clear
Preston Guild Hall	Phase 1	Preston	Events	1970	1991	High	Central warm air	Daily	5	3		4	1 VWR	Concrete	Bituthene	50 mm	50 mm	Vapour sealed	Top and bottom	Outside
	Phase 2				1992								VWR							Outside
	Phase 3/4				1993	Phase 4 1995							VWR							Outside
Salisbury Cathedral	Typical	Salisbury	Cathedral	>500 yrs	Varies	Cathedral	HW radiators?	Continuous	3	3		3	2 Vaulted	Vault	None	None	Walkable	Via 100 mm hole	Small windows	Outside
	SW corner																			Outside
St Cross, Winchester	Typical	Winchester	Church	>500 yrs	1880s	Daily serv	Direct gas	Short bursts daily	3	2		3	3 Vaulted	Stone	None	None	Walkable	Via doors etc	Small windows etc	Both
	Tower								2	1		5	3 Cold	Timber	None	None	Walkable	Via floorboards	Belfry openings	Mostly O/S
St Mary's, Hadleigh	New VWF	Suffolk	V church	>500 yrs	1988	V church	Gas plaque+pew	Occasional	4?	2?		3?	2 VWR	NA	Polythene	50 mm?	50 mm?	Vapour check	Side-to-side	Outside
St Mary's, Stoke-by-N	Typical	Suffolk	V church	>500 yrs	1967	V church	Electric pew	Occasional	3	2?		2?	4 Direct	None	None	None	None	Complete	Outlet via gaps only	Inside
St Mary's, Stratford	S aisle	Suffolk	V church	>500 yrs	1986	V church	Warm air+pew	Occasional	4?	2?		3?	2 U'drawn	Timber	None	None	Shallow	Via cracks/gaps	Via slate area	Inside?

62   English Heritage Research Transactions Volume 1

Table 4. Lead roofs inspected in the UK (continued).

NAME	Variant	SUBSTRATE PITCH: 1=<3 2=<6 3=<15 4=<30 5=>30 6=90°	LEAD TYPE	CODE VERT JOINTS	HORIZ JOINTS	SPLASHLAPS	UNDERLAYS Lower layer	UNDERLAYS Upper layer (if any)	UNDERSIDE CORROSION ASSESSMENT Over boards	Over gaps	At laps or steps	At rolls	COMMENTS	POSSIBLE FUTURE WORK
Brightling Observatory	Typical	2 S/W	Mill?	6 B/R	Step	Yes	None		Slight	Some streaks	Slight	Modest	Some corrosion where sealants used	None, currently on market
Caerhays Castle	Typical	2 S/W	Cast	8 B/R	Step	Loose	None		Slight	Passivated	Some	Slight	Brown topside patina in warm damp climate	Phase I re-roofing with chalk just completed
	Elm boards		Cast	8 B/R	Step	Loose	None		In some areas	Some	Slight	Slight	Chalk coating tests give some protection	Phase I re-roofing with chalk just completed
Cathedral in North West	Existing	5 S/W	Cast	6 H/R	Lap	No	None		Some	Not inspected >>>>>>>>>>>>>>>>	Substantial	Substantial	Water ingress main cause?	Test chalk samples in corroded areas?
	New VWR		Mill		H/R	Lap	No	Geotextile		In "dead" areas	Significant	Some	Slight	Most in flat areas above upper vents	Test chalk samples in corroded areas?
Church in Bucks	Aisle	2 S/W	Cast	6 B/R	Step	Small	None		Slight	Significant	Limited	Limited	Slippage shows past corrosion above gaps	Tests and monitoring continue
	Nave	2 Gap S/W	Cast	6 H/R	Lap	No	None		Significant	Varies	Largely passivated		Pressurised direct gas-fired heating might be increasing corrosion now	
Church in Northants	Nave	2 S/W	Cast	7 H/R	Lap	No	420 B paper?		Bright	Bright	Some	Minimal	Particularly well heated and ventilated	Check autumn 1996
Church near Sheffield	Existing	2 S/W	Cast	7 B/R	Lap	No	420 B paper		Severe: failing	Worst at laps	Yes	Yes	Drying-out problems with constant heating	Chapel was very damp: sealed+gas htrs!
	New VWR	3 S/W	Mill	7 B/R	Lap	Yes			Little	Some at edge	Slight	Slight	Vapour check not well sealed at edge	Lead stolen. Improve details when replacing
Church in Shropshire	Aisle	1 S/W	Mill	6 B/R	Step	Tight	Insul bd 13 mm	420 B paper	N/A	N/A	Substantial	Substantial	Thermal pumping and ingress likely	Check performance of closed cell insulation
Church 1 in Yorkshire	Nave	3 Oak	Cast?	7 H/R	Lap	No	Hot bitumen	420 B paper	Significant	Significant	Severe	Severe: failing	Bad just above lap, likely acid attack	Chalk gives useful protection to samples
Church 2 in Yorkshire	All	3 T&G S/W	Mill	6 H/R	Lap	No	None		Some in middle	Ditto	Slight	Slight	Exposed site, top lap has more corrosion	Consider installing small dehumidifier
Church 3 in Yorkshire	Chancel	3 Oak	Mill?	7 B/R	Lap	Yes	None		Substantial	Passivated	Passivated	Slight	Oak very wet when inspected March 1994	Test remedial underlays
	Aisle	3 S/W	Mill	7 B/R	Lap	Yes	None		Minimal	Minimal	Minimal	Minimal	Only one sheet lifted, much drier	
Civic building	Typical	4 S/W	Mill	7 B/R	Lap	No	None (diagonal boards)		Not normally	No	Slight	No	Excellent where vapour check (VC) good	Check autumn 1996
	Leaky VCL	4 S/W	Mill	7 B/R	Lap	No	None (diagonal boards)		Not normally	Yes	Slight	No	Poor where ventilation or VC faulty	Check autumn 1996
Donnington Castle	Existing	2 Oak	Mill	7 B/R	Step	Yes	Hardboard		Severe: failing	Less	Steps passiv	Yes	Large set of test samples in place	Revisit to check samples.
Educational building	Typical	3 Plywood	Mill	7 B/R	Lap	Yes	Insul board	Plywood	Extensive	Extensive	Extensive	Varies	Passivated over adjacent woodwool slabs, though slightly corroded at some joints	
	Facia	6 Plywood	Mill	7 Welt	None		None		Disastrous	N/A	Disastrous	N/A	But passivated nearby!. Welts trap rain	Inspect renewed roof
Hampton Court	Gt Hall	5 Oak	Mill?	7 B/R	Lap	Small	Hardboard		Severe in places	N/A	Severe	Severe	High acid content in hardboard	Monitoring and testing continues
Manchester Cathedral	Aisle	3 Ply?	Mill	7 B/R	Step	No	420 B paper		Slight corrosion	N/A	Some	Not known	Monitored in 1988-89	Inspect autumn 1966
Mansion in Bucks	Original	2 S/W	Mill?	7 B/R	Step	Yes	Hairfelt		Extensive	Passivated	Passivated	Limited	Many nosings have been repaired	Chalk treatment looks promising: continue
	New	2 S/W	Mill	7 B/R	Step	Yes	Geotextile		Minimal	Substantial	Substantial	Substantial	Fish tails rising from gaps into rolls	Chalk treatment looks promising: continue
	Separate	2 Plywood	Mill	7 B/R	Step	Yes	420 B paper		Slight	N/A	Substantial	Some at s/laps	Polythene seals poor but A/C not humidifier	Test coatings including chalk
Mansion in Derbyshire	Existing	2 S/W	Milled	8 B/R	Step	Yes	Erskine's felt		Patinated	Patinated	Slight	Slight	Buffering effect seems helpful here	Monitoring and testing continues
	Renewed	2 S/W	Cast	8 B/R	Step	Yes	None		Spread from gaps	Significant	Some	Slight	Reduced buffering has caused corrosion	Test coatings and blocked ventilation
Mansion in Dorset	West end	2 S/W	Mill	7 B/R	Lap	Yes	Erskine's felt		Initial corrosion	More now	Some at s/laps	Some at s/laps	Samples of treated underlays left	Return to inspect in 1997
	East end	2 S/W	Mill	7 B/R	Lap	Yes	Erskine's felt		Initial corrosion	Less	Slight	Some at s/laps	Samples of treated underlays left	Return to inspect in 1997
	Mansard	5 Gap S/W	Mill	7 B/R	Lap	Yes	None (diagonal boards)		Little	Some	Slight	Not inspected >>>>>>>>		Return to inspect in 1997
Mansion in Northants	Renewed	2 Plywood	Mill	7 B/R?	Step	???	Bldg paper	Geotextile	Not known				Not visited since re-roofing	Inspect renewed roof
	Chapel	1 S/W	Mill	6 B/R	Lap	Tight	Erskine's felt		Some	Some	Local	Some	Lap mean: some ingress suspected	Check other locations
Metal store		1 Plywood	Cast	7 B/R	Step	Yes	Insul bd 13 mm	420 B paper	Slight	N/A	Heavy	Severe	Step at bottom (into gutter) had no splashlap and was not corroded	
Museum		2 S/W	Cast	7 B/R	Step	Yes	Various B papers		Slight	N/A	Slight	Locally heavy	V well H&V roofspace; foil faced insulating board under softwood deck	Investigate origin of acid causing localised roll corrosion
Norwich Cathedral	Choir sch	3 Oak	Cast?	8 H/R	Lap	No	None		Severe: failing	Similar	Severe	Severe	Failing, many repairs	Consider installing coated test samples
	Refectory	2 Oak	Cast?	8 H/R	Lap	No			Only inspected from above				No signs of failure from above but no sheets lifted	
Preston Guild Hall	Phase 1	4 S/W	Cast	7 B/R	Lap	No	420 B paper		None	None	Slight	Slight	Some leaching by organic solvent	
	Phase 2	4 S/W	DM	7 B/R	Lap	No	234 B paper		Minimal	None	Slight	Minimal	Slight yellowish corrosion over tanalised boards, not serious	
	Phase 3/4	4 S/W	Cast	7 B/R	Lap	No	234 B paper		None	None	Slight	Minimal	Performance seems good	Check
Salisbury Cathedral	Typical	5 Gap S/W	Cast	8 S/S	Lap	No	None		Little	Slight, varies	Not known		Early gap corrosion sometimes occurs but settles down	
	SW corner		Cast	8 S/S	Lap		B paper 420						Small part of roof only, some problems but origin unclear	
St Cross, Winchester	Typical	5 Gap S/W	Cast	8 B/R	Lap	No	None		Minimal	Passivated	Slight	Slight	Fresh samples in church corrode at gaps	Revisit and install remedial test samples?
	Tower	3 S/W	Mill	7 B/R	Lap	No	None		Slight	Some	Significant			
St Mary's, Hadleigh	New VWR	4 S/W	Mill	7 B/R	Lap	Loose	Geotextile		Significant	Significant	Minimal	Some	ULC worst in dead spot, centre bottom	
St Mary's, Stoke-by-N	Typical	2 Oak	Mill	7 B/R	Step	No	Bitumen/Erskin	420 B paper	Small	N/A	Significant	Some	Water ingress path at foot of hollow roll	Inspect test samples?
St Mary's, Stratford	S aisle	2 S/W	Mill	7 B/R	Step	Yes	Slaters felt	Breather paper	Some	N/A	Occasional	Some at s/laps	Some ULC patterning with ridges and furrows of breather paper	

^ B/R=Batten Roll; H/R=Hollow Roll; S/S=Standing Seam.

corrosive elm or gaps but there was some corrosion where chalk-loaded geotextile underlay was used underneath.

Conclusions: Demonstrates the usefulness of a chalk coating retained in place, but the benefits of chalk-coated geotextile are questionable. Confirms passivation in some humid environments. A section has been re-roofed more or less to the original specification with the sheets reduced in width and a chalk coating applied. This will be examined in 1997.

- *Cathedral in North West: observations, plus some monitoring by BRE*
  Reasonably heated. Various roofs upgraded to ventilated warm roof specification when re-laid, with softwood gap-boarding and geotextile. BRE monitoring shows no significant moisture entering the air gap from inside the building. Nevertheless, some cosmetic corrosion was found, particularly above the outlet vents on lean-to areas and at laps.
  Conclusions: Corrosion seen in the laps is almost certainly caused by distilled rainwater. Important to have through-ventilation in a ventilated warm roof, with no dead spots. Geotextile might not be the best underlay. Comparison tests are desirable.

- *Chester Cathedral, Cheshire: observation*
  Several lead roofs here have recently been replaced using DM lead, but concern was expressed at their appropriateness and appearance. The material has one rippled face and one flat face: its appearance is normally satisfactory with the flat face outermost; see Preston Guild Hall.
  Conclusions: If DM lead is used, for appearance it should have the flat side upwards. It would be worth undertaking studies of whether any patterning effects are associated with the ripples in corrosive situations (as puckered building paper sometimes has had on flat lead).

- *Donnington Castle, Berkshire: observation, monitoring and tests*
  No heating. Lead laid in the 1950s over new oak with hardboard underlay is badly corroded over much of its area. High acetate contents (600 ppm) in the hardboard, in which acetate seems to accumulate. Less corrosion (and some passivation) where there is softwood (rather than oak) under the lead, in spite of similar hardboard acetate content, probably the result of hygroscopic buffering by the softwood. Similar, but less marked, effect over rafters and purlins, and some passivation in parts of rolls. Wet gutters uncorroded, possibly owing to the effect of carbonated concrete. Roof used by RTL in 1994–6 for full scale tests.
  Conclusions: Beware accumulation of organic acids in certain materials. Chemical effects of acetate seem to be greatly influenced by local hygrothermal conditions. Recent tests with chalk coatings and underlays are promising.

- *Educational building in Cambridge: observation and tests*
  This roof on a late 1960s building had failed badly, owing to a combination of underside corrosion, water ingress via a fascia welt detail which had become a water-trap, and with cracking elsewhere from thermal movement. There may also have been some condensation. When opened up, the corrosion was found to be restricted to lead either laid on plywood, or in the distillation zone above areas of damp plywood. Although corrosion was very severe, with the lead paper-thin in places, passivated areas were found in close proximity to corroded ones, even over the plywood. Sharp boundaries between the two states have also been seen in particular at Donnington Castle, Hampton Court and a church near Sheffield. Lead over woodwool cement slabs here was also well passivated, in spite of evidence that it had been subject to condensation from time to time.
  Conclusions: An example of the severe corrosion caused by the hydrolysis products of damp plywood. After a hot dry month, the plywood here was also found to be very damp in the middle although its surfaces were dry. Conversely, the carbonated cement in the woodwool appeared to have had a passivating effect.

- *The Great Hall, Hampton Court Palace, Surrey: observation, monitoring and tests*
  The steeply-pitched roofing of the mid-1950s has a hardboard underlay (now with a very high acetate content) over oak decking, similar to Donnington Castle. However, unlike Donnington the Great Hall has background (underfloor) heating and also has bitumen-cored building paper under the hardboard. Both north- and south-facing slopes are badly corroded, particularly around the edges of the sheets and in places where the lead has arched away from the substrate owing to constrained slippage (the lead is nailed to the rolls from top to bottom) and thermal movement. Such patterning appears to be widespread where organic acids are involved, and may be related to distillation across the air gap, greater local access of carbon dioxide to regenerate acetic and formic acids, and electrochemical differences. The building paper also appears to have helped to trap ingressed moisture in the hardboard layer. Although the internal environment is relatively dry, removing the hardboard and building paper in a sample area led to visible evidence of condensation. This may result from the sustained egress of moist air via natural buoyancy in this tall, single-volume building which has no openable windows and a timber ceiling with joints readily permeable to the passage of air.
  Conclusions: Confirms the particular acidity problems of hardboard, probably exacerbated by nearby oak. Suggests that the building paper was an ineffective barrier and may have made things worse by trapping moisture. Illustrates that corroded and passivated areas can be in close proximity, even in a highly acid environment.

- *Mansion in Northamptonshire: observation only*
  This house was visited while lead was about to be installed on geotextile over a plywood deck. While it was too late to change this, to reduce the risk of problems better air and vapour sealing of the ceiling and better ventilation of the roof space was recommended, together with building paper under the geotextile. The renewed roof will soon be inspected.

- *Church in Shropshire: observation, tests and monitoring*
  In the early 1980s, the south aisle roof was renewed, as recommended at the time, with a vapour control layer bedded in hot bitumen over plywood, 12 mm wood fibre insulation board and bitumen-cored building paper under the lead. Five years later the insulation board was found to be wet and the underside of the lead corroded, particularly at the perimeter near laps and rolls. At the time the problem was attributed to a faulty vapour control layer, and possibly water ingress at the steps which did not have the specified anti-capillary grooves (although the Lead Sheet Manual does not require them for the 50 mm steps used here, but only for shallower ones). Battens at the bottom of the sheets above a step could also trap moisture at the foot of the insulation boards.
  Environmental monitoring is still in progress with tests on moisture-resistant non-porous insulation, and different anti-capillary and water run-off systems but none seems to have been successful. Results to date suggest that the church is well-heated and relatively dry (reducing the risk of condensation) and that vapour control layer is effective. It is therefore likely that the moisture in the insulation board originates from outside, not inside. This moisture is effectively trapped: it did not even dry out in the prolonged hot dry summer of 1995. Water ingress paths have been identified via poor rendering above the top flashings and sub-atmospheric pressure ('thermal pumping') at the foot of the rolls just above the steps.
  In the laps of the more steeply-pitched chapel roof, there is also some local evidence of water ingress (probably by thermal pumping, but it could be a wind effect) via nail holes which have become elongated owing to movement against the soft insulation board. However in the chapel bays inspected there was little underside corrosion.
  Conclusions: Since 1986 'warm' lead roofs like this have not been recommended. Low-pitch roofs with splashlaps (such as this) are particularly susceptible to thermal pumping because rainwater retained in the splashlap creates a water seal which permits high negative pressures to develop under the lead when it cools, and a reservoir of water to be drawn in.

- *Mansion in Dorset: observation, monitoring and tests*
  When restored in the early 1980s, roof space ventilation was improved using ventilated access hatches and 15 mm copper tubes at regular intervals around the eaves. The lead was laid directly on softwood gap-boarding on the mansards and close boarding with Erskine's felt underlay on the top. At the first quinquennial inspection, underside corrosion of the flat-roofed areas was virtually universal, though generally thin, compact and quite protective. Studies indicated that most of this had formed early in the life of the lead. In sample areas where the lead surface was cleaned, new corrosion only occurred above the gaps between the boards over the occupied flat on the top floor (in which additional moisture was generated by the occupants and their activities) and not above the exhibition rooms (which today have conservation heating). The additional roof ventilation openings did not seem to have been particularly helpful, often working as outlets for air rising from within the building. In some places where felt had been omitted the lead was covered with a loose, friable, dusty or flaky corrosion product: when cleaned off, no further corrosion occurred, indicating that this probably resulted from initial contact with damp and/or freshly preservative-treated wood. In active corrosion sites, corrosion could be reduced by applying linseed oil or patination oil to the underside of the lead: this was only fully effective if the oil was given sufficient time to cure before laying. Patination oil was generally better in practice because it cured more quickly. Some corrosion was also found on the outer parts of the rolls towards the splashlaps: this was thought to be from distillation of rainwater trapped in the splashlap, and has since been found on many other sites with low-pitched roofs.
  Conclusions: Visible corrosion may have occurred early in the life of a roof and might no longer be active, so it is important to check before taking action which could be unnecessary. Linseed oil, which was sometimes applied to lead in the past (both in some rolling mills and on site), could have conferred some resistance to underside corrosion, at least initially. Pre-coating with patination oil may be helpful (but chalk treatment may be preferable, subject to further tests). Additional roof space ventilation may not always be effective or necessary.

- *Manchester Cathedral: observation, tests and monitoring*
  The original test site for electrochemical monitoring of condensation corrosion of lead. The tests showed that on a sample arranged to be susceptible to condensation and corrosion, most of the corrosion occurred in the periods during which the condensate was drying out. The renewed nave roof, laid on building paper on softwood decking, only had a small amount of underside corrosion.
  Conclusions: Important initial site. Worth re-visiting to check the current condition of the lead.

- *Hotel in the Midlands: observation*
  The mansard roofs on this 1970s building were formed from Codes 4 and 5 lead sheet bonded to plywood and with a central metal clip driven into the wood and welded to the lead. Thermally-induced cracking had occurred (especially on the south and

east faces), with ridges formed by compressive expansion turning into cracks and ingress of moisture then causing corrosion damage. Even before this, some condensation and corrosion might well have happened. The lead was both over-sized and over-fixed, being secured to the plywood right around the edge and with clips in the middle. The corrosion had been greatly accelerated by acetic and formic acids from the plywood.

Conclusions: Some suppliers have argued that bonded lead can be used in larger panel sizes than recommended in freely-suspended situations because the loads are spread and buckling is restricted. However, we have found no firm evidence for this and bonded sheets are not covered by British Standards and LSA recommendations. Expansion and over-fixing was a bigger problem here (at least initially) than corrosion. It is also possible that, even using LSA size and fixing recommendations, bonded lead might ultimately suffer tensile fatigue because the adhesive (which stiffens as it cools) would restrict thermal movement and particularly contraction. Great caution must also be exercised in the choice of manufactured timber-based substrate boards owing to their potential acidity, particularly in environments in which there is any risk of dampness.

- *Norwich Cathedral cloisters: observation*
The quadrangle of cloisters has first floor rooms above, containing a choir school, a library, a restaurant and an audio-visual room, all with oak substrate boarding of some antiquity, which also forms the ceiling. New lead was laid in phases during the 1950s and 1960s. A lowered ceiling with vapour control layer and roof space ventilation was added above the restaurant only. The choir school roof, immediately beside the cathedral and subject to egress of moist air up the connecting stair, is the earliest and the worst corroded. Owing to the hollow roll construction, inspection of the underside of the lead was difficult, and sheets were only lifted on the worst-corroded part. Corrosion here was widespread, and worst along the joints between the boards and around the edges: this is characteristic of situations in which organic acids are involved. An inspection from above and below suggested that there was less corrosion in the newer roofs above the library (which was drier at the time) and even less above the restaurant. The roof of the audio-visual room (adjacent to the restaurant, affected by moisture emerging from it and without a lowered vapour-checked ceiling) was more suspect.
Conclusions: Oak, even if well-seasoned, may still promote underside corrosion. These roofs were inspected several years ago. A fresh and more detailed inspection, plus possible tests, would be desirable.

- *Preston Guild Hall, Lancashire: observation and tests*
The roof has recently been replaced with ventilated warm roof construction. Ventilation rates have been checked in relation to wind direction and solar heating, and the underside of the lead inspected. The Phase 1 roof (sand cast Code 8 lead) showed some adhesion between the lead and the substrate (Sisalkraft 420), owing to leaching of bitumen from the building paper's core by residual solvents for the wood preservatives. The Phase 2 roof (DM Code 8 lead on Sisalkraft 234 plain reinforced building paper) had been installed the correct way (flat side of the lead up). An initial yellow corrosion product indicated some reaction with the water-based softwood preservatives, or with the damp treated wood, but this corrosion appears to be cosmetic only. In the recent Phases 3 and 4, Code 8 sand-cast lead was used, with Sisalkraft 234 and a requirement that the treated timber should be dry. These roofs have not yet been inspected.
Conclusions: Well-detailed ventilated warm roofs appear to work well but there need to be precautions against damp or freshly preservative-treated substrate boarding.

- *Civic building in Hertfordshire: observation, tests and short-term monitoring*
A lead roof laid on a plywood deck with insulation underneath and poor vapour check details failed within a few years. Its replacement with a ventilated warm roof performed well generally, but with condensation and underside corrosion in a few places. In some of these, the airspace did not have through-ventilation from eaves to ridge. Where there was through-ventilation, tracer gas tests revealed that corrosion occurred in the bays in which there were faults in the sealing of the vapour control layer. These included junctions to brickwork and penetrations such as roof windows, where moist air and water vapour from inside the building could rise into the ventilated void.
Conclusions: The original roof was replaced before this project started and we have no records of it. However, the research indicates that it could only have failed as fast as it did if condensed and trapped moisture had activated the acids in the plywood. While the best parts of the new roof demonstrate the effectiveness of ventilated warm roof details, the weak spots makes it clear that the whole of the airspace must be ventilated by a through-flow of outside air (with no dead spots). The vapour control layer must also be meticulously detailed and jointed so that it is both vapour-resistant and airtight: ventilated air spaces cannot be guaranteed to bear away any moist air or water vapour with no ill effects. Adding a layer of plain reinforced building paper (such as Sisalkraft 234) under the lead can be helpful (see Preston Guild Hall), as can chalk treatments, but these need further testing.

- *Church near Sheffield: observation only*
A damp and poorly ventilated church, with ventilation further reduced in the north-east chapel owing to a modern enclosure. Although it had been intermittently heated, some months before our visit there had

been a change to continuous heating. This had increased evaporation from the walls and caused some efflorescence, but had not dried them out, partly owing to the poor ventilation and possibly to abundant sources of moisture. The raised internal dewpoint had also made it very wet under the lead roofing. The chapel roof had probably been wet before, owing to its enclosure and past use of flueless bottled gas heaters: it had corroded through in places, and was being replaced by ventilated construction. The decking was softwood (with lapped bitumen-cored building paper over), the building paper was very wet (though not decayed) and there was evidence of water vapour (and acid?) egress via the laps. The oak rafters and purlins underneath appear to have contributed to the corrosion, and had locally corroded the lead in the gutters near the exposed rafter ends.

Conclusions: Dampness and insufficient ventilation had exacerbated any problems. The increased heating, without attention to ventilation and drying-out, had made matters worse. The renewed chapel roof will soon be inspected.

- *St Mary's, Hadleigh, Suffolk: observation only*
Eleven bays at the north-east corner of the nave were re-covered in 1988 with an early version of the ventilated warm roof, with lead laid on geotextile over softwood boards with penny gaps. Unusually, the ventilation was not from eaves to ridge, but from side to side, with two ridge-like vents running up the pitch. The lead here shows some underside corrosion, with a white product which is friable and non-protective. The corrosion is greatest in the middle and particularly at the bottom, where ventilation is likely to be poorest. There is also evidence of condensation having trickled down from time to time. The effectiveness of the vapour control layer is not known.

Conclusions: This further supports the requirement for meticulous detailing and 100% through-ventilation in a ventilated warm roof. It also seems that the use of geotextile as an underlay may have increased the amount of corrosion, at least initially.

- *Church in Northamptonshire: observation only*
This was visited two years after the nave had been re-leaded, with hollow rolls and bitumen-cored building paper over the existing softwood boarding (there is a wooden ceiling immediately below this). The underside of the lead panel lifted was in very good condition. This was thought to be because the church was continuously heated, well-ventilated (with the churchwarden providing additional ventilation by opening doors and windows on warm, dry afternoons) and consequently reasonably dry.

Conclusions: Although such a detail is potentially at risk, it appears to have been protected by the benign environment and an assiduous churchwarden. Changes to the environment could alter this situation, and indeed internal staining indicated that there had been moisture problems in the past. A re-visit is planned.

- *St Mary's Church, Stoke-by-Nayland, Suffolk: observation only*
This church has oak ceilings throughout. All roofs were reportedly re-covered in 1967 with Code 6 milled lead, laid on bitumen-cored building paper (probably Sisalkraft 420), with what looks like Erskine's felt with a high bitumen loading under that. The underside was difficult to inspect owing to the hollow rolls, but on the south aisle it appeared to be passivated, except for some faint stripes of corrosion. The nave roof, however, was more corroded, though it was not clear whether the oak or rainwater ingress was the prime cause (a possible leakage route was identified beneath the nosings). If moist air and acid egress was to blame, to find more corrosion over the nave than the aisles is not unusual, because both wind and natural buoyancy forces tend to cause more air to leave at the top.

Conclusions: An interesting variant on the double-layer theme (where the less permeable layer is normally underneath) and one which has performed reasonably well. However, the detail does not have a clean bill of health, and small amounts of trapped moisture, whether from ingress or condensation, are still problematic.

- *St Mary's Church, Stratford St Mary, Suffolk: observation only*
The south aisle of this church was re-leaded in 1986, with steps formed and new plywood decking over, covered with an impervious layer of slater's felt underneath and breather paper on top. Joints in the slater's felt had been sealed with hot bitumen and some of this had partly been absorbed into the breather paper, which nevertheless still formed a good slip layer. In spite of the plywood (and an oak ceiling and oak structure underneath), the underside of the lead was generally in good condition, although in a few places there had been rainwater ingress and underside lead corrosion (although still cosmetic) had started. As at the Museum building in London there was also some corrosion above areas where the building paper had become impregnated with the bitumen, and no longer had an absorbent upper surface. As at the house in Dorset there was also some corrosion in the rolls above the splashlaps. There had also been some movement in the felt which was showing signs of cracking in places: perhaps this would not have happened had it been fully bonded to the deck.

Conclusions: While confirming the potential for double-layer underlays, evidence from this site supports other tests which suggest a vulnerability to corrosion by any ingressed and trapped rainwater.

- *Church 2 in Yorkshire: observation and monitoring*
A very damp village church in the Yorkshire Dales. The pitched roof was renewed in 1990 directly over the existing tongued-and-grooved ceiling/substrate boarding. At the same time the central heating was

changed from column radiators to under-pew skirting heating. Occasional dripping condensation was then reported: initially thought to be emerging from under the lead, but calculations indicated that it was surface condensation after still, clear nights in the harsh microclimate. Droplet formation is concentrated beneath the tips of the nails used to fix the lead: these go nearly all the way through the ceiling boards, and today's use of longer, stouter, highly-conductive copper nails probably exacerbated the problem. The new heating, although more efficient, may well have made condensation more likely by heating the underside of the roof less: column radiators often make the air highly stratified, while with the under-pew system the measured temperature gradient was small. The church may also have become damper generally for several reasons. The average ventilation rate measured by BRE was 0.7 air changes per hour, somewhat above the CIBSE Guide's rule of thumb of 0.5 ac/h for a small church.
Conclusions: Detailed changes to heating and to lead fixings can significantly affect the outcome in marginal circumstances, and in this particular climate the condensation risk is high too. So far the problems are mainly the dripping condensation, but there is also some underside corrosion and RTL are undertaking tests. Steps should be taken to reduce moisture levels in the church, preferably at source by attention to pointing and rainwater systems, and additionally by heating or dehumidification. A dehumidifier was tested for a few weeks and had a visible effect although part of this could have been related to the unusually dry 1995–6 winter. However its operating cost of about £2.50 per winter day was deemed high by the church.

- *Church 3 in Yorkshire*
  Another very typical church with the same boarding forming the ceiling and supporting the lead. The chancel roof has an oak ceiling, over which the lead was badly corroded, particularly above the boards (there was some passivation over the gaps). Over the softwood boards in the aisle, however, the lead was in reasonable condition. Although subjected to the same internal atmosphere, the oak boards were very wet and the softwood boards were not, a consequence of their different hygroscopic properties.
  Conclusions: The difference in corrosivity of different woods is related to both their physical and their chemical properties. This is a potentially useful site for testing remedial underlay and/or coating specifications for oak.

- *Salisbury Cathedral, Wiltshire: observation, testing and monitoring*
  All roofs inspected were relatively good condition. Temperature and humidity monitoring was carried out in several places. The buffering effects of the large volumes of air and hygroscopic material (particularly timber) in the roof spaces are thought to have helped to protect the lead from severe underside corrosion.
  Conclusions: Such roofs appear to be protected by three mechanisms:
  - moisture-stabilization by large volumes of absorbent material
  - a degree of isolation from the atmosphere in the building underneath
  - passivation of the lead by moisture which emerges from the wood when the sun heats the lead and the roof space.

  Good ventilation via the gap-boarding also helps, where the roof space is relatively dry. However, the mechanisms and their interactions are not yet completely understood.

- *Mansion in Derbyshire: observation, testing and monitoring*
  The lead here covers both the top and the sides of some mansard roofs. During refurbishment of one roof, the opportunity was taken to improve outside air ventilation (using 'cold' roof principles) and to add fire barriers. At the same time, to avoid the ingress of driving rain and snow, copper flaps were added which closed when air velocities through the ventilators were high.
- The new roof shows more corrosion than a similar existing roof, which was also not so well ventilated. There are four likely reasons for this:

  - moist air and water vapour are entering the roof void from below, and with the lower roof space temperature the timbers have become moister and the likelihood of corrosive conditions has increased
  - the flaps are not working as intended and moisture is being trapped
  - the additional ventilation has undermined the buffering mechanisms which occur, for example at Salisbury Cathedral. This could have increased the number of evaporation/condensation cycles and at the same time have made it more difficult for the lead to self-passivate spontaneously in warm, sunny weather
  - unfortunate starting conditions, leading to unprotective initial corrosion.

  Monitoring is currently being undertaken and shows that the roof space is very cold, the local microclimate very damp and the potential for condensation is high. Future tests are planned with variable amounts of ventilation.
  Conclusions: Increased ventilation is not necessarily desirable unless one can attain ideal 'cold roof' conditions, in which there is a highly effective air and vapour seal at ceiling level and 100% outside air through-ventilation of the roof space. This is very difficult to achieve in any existing building, let alone a historic one, other than by means of well-detailed ventilated warm roof construction.

- *Metal store in London: observation*
  A brief inspection here has revealed heavy corrosion around the perimeter of lead laid on Erskine's felt on plywood. This is characteristic of attack by acetic and/or formic acids. The environmental conditions have not yet been characterized, but being over a metal store, they are unlikely to be particularly humid.
  Conclusion: Beware plywood in all but the driest conditions.

- *Museum in London: observation*
  The main roof was replaced in 1966–7 with a completely new steel structure, from which the original oak beams, purlins and boarded ceilings are suspended. The lead is laid on good quality 30 mm softwood boarding with penny gaps, with an intervening layer of building papers of various grades, sometimes bitumen-cored and sometimes not. The main pitches of the roof and upstands (but not the gutter soles) have a layer of aluminium foil-faced insulation board, with the foil face upwards (touching the underside of the decking boards). The roof space is ventilated by warm, relatively dry air rising from inside the building via gaps between the ceiling boards and out through louvres in the upstands to the valley gutters.[34] The underside of the lead is generally in very good condition, although with some traces of corrosion in places, either associated with water ingress, joints in the insulation board and places in which bitumen has leached out of the building paper above knots and resin pockets. Corrosion on the outer parts of some of the rolls, above the splashlaps, is quite severe and is being investigated. There is a high organic acid content in these locations.
  Conclusions: The main reason why most of this roof has performed well is because the building and the roof space is relatively dry, well-heated and well-ventilated. Even though there are condensation risks at times, in warmer conditions the constant, warm ventilation will have helped the timber to dry out well, making it able to absorb considerable amounts of moisture during adverse conditions. The main roof construction with the aluminium foil vapour control layer (which was still in excellent condition where we inspected it) approximates to that of a 'warm' roof and might have been expected to be susceptible to thermal pumping. We suspect that this did not happen in practice owing to:

  - the relatively small volume of air which is trapped in the timber boards, at least in relation to open-cell insulating materials
  - pressure-relief via the joints in the insulation boards, which are not taped.

  However, the entrapment of the timber may be the reason for the high organic acid content in the slashlaps. There has recently been a proposal to humidify the museum to improve conditions for exhibition display. We have expressed strong reservations about this, and we understand that humidification may now be restricted to some basement areas.

- *Mansion in Buckinghamshire: observation, testing and some monitoring*
  An early twentieth-century mansion which had a variety of uses before converted to its current use as an office and training centre. The roofs have void spaces which are not deliberately ventilated. The lead, thought to be largely original, is mostly laid on hair felt over softwood boarding. The original lead is usually significantly corroded above the boards, with a compact but sometimes flaky layer of yellowish corrosion product, but the corrosion product is relatively thin (5% or less of total thickness) and has not affected the life of the roof. Above the gaps between the boards, the original lead is passivated. Many nosings have been repaired and a few sheets replaced.
  A small number of sheets were renewed in the late 1980s and laid over geotextile in October. These immediately began to corrode above the gaps between the boards. Since then the corrosion has continued, and also spread into the rolls. Sample tests however, showed that lead laid in the early summer was much less susceptible to this type of corrosion. Environmental monitoring indicated that the building was relatively dry.
  Conclusions: These were the first tests to show that, at least in marginal situations, the time of laying and the underlays used might have substantial effects upon corrosion behaviour. It would be worth returning to this site to undertake more inspections and tests, including tests of specifications (such as chalk treatments and underlays) which the research now suggests could be really helpful in situations such as this.

ENDNOTES

1. The original paper recommending ventilated warm roofs (Murdoch 1987) and current guidance (LSA 1993a) draws attention to some of these detailed issues. However, experience in practice suggests that there would be no harm in underscoring these more strongly, and probably including some drawn details of do's and dont's, as in the BRE publication *Thermal insulation: avoiding risks* (1994). Murdoch (ibid) also stated that the ventilated layer would help to disperse any moisture that did penetrate the vapour barrier: while this is correct, site experience indicates that any weakness to the passage of water vapour, and in particular moist air, may initiate underside corrosion.
2. Solubility of the oxide and hydroxide is lowest at pH 9.5 and increases rapidly with rising acidity or alkalinity (Pourbaix et al 1966). Protection will therefore be best at pHs between about 8 and 11.
3. In the 1970s patination oil was developed to control initial weathering and avoid white staining of brickwork subject to run-off. Its effect is primarily physical, as a barrier layer to keep water and lead apart while the lead has time to develop its own patina underneath under the influence of light and chemicals able to diffuse through the oil layer.
4. The higher atmospheric concentrations of sulphur in the age of coal-burning may be significant here.

5. Underside corrosion is not restricted to lead. Some of the physical principles also apply to other roofing materials, particularly zinc and aluminium (Farrell et al 1992).

6. Excess moisture and condensation may also affect the roof structure, whether or not the covering material is resistant to underside corrosion. As a general rule, decay fungi only become active if wood has a moisture content above 20%. Wood-boring insects also prefer moist timber, typically over 15%: at lower moisture levels their activity diminishes and below about 10% they cannot survive (Oxley & Gobert 1994, Ridout 1995).

7. However, on a roof, if this hard, compact film becomes too thick, stresses from thermal movement of the lead cause it to crack and spall, leading to multiple layers of a dense, flaky corrosion product.

8. If so, since lead oxide is slightly soluble in water it may help to explain why passive films have only a limited life in continuously condensing conditions, or after many condensation/evaporation cycles in the test rig, as these would tend to wash away some of the oxide. With more occasional condensation, dissolved oxide might crystallize out again, and in any event there will be more time for self-repair to occur. It is also possible that the oxide is covered by a less soluble layer (for example containing carbonate or even sulphate), which confers increased initial corrosion resistance.

9. Some modern sealants, including some silicones, also emit acids when curing, with similar corrosive effects.

10. However, the oak used in RTL's external test facility was found to have a pH of 4.2, putting it in the high rather than the severe category (RTL 1995, Report 6)

11. With the exception of wood fibre insulation board which Hill found to be slightly less corrosive than Swedish whitewood and redwood, and (just) the least corrosive of all the samples tested (Hill 1982).

12. Museums have also found that manufactured boards tend to be highly variable in their corrosive effects, not only by product type, manufacturer and source, but also from batch to batch (Tennent, Tate & Cannon 1993).

13. To avoid interstitial condensation, this inner layer of insulation must not be too thick in relation to the outer one.

14. Ventilated warm roofs have a buildability advantage in that the AVCL can be placed rapidly and, if suitably detailed, act as temporary weatherproofing. However, to avoid initial moisture and corrosion the insulation and the supporting structure for the lead should nevertheless be protected from rain during construction.

15. *Architecture with Rheinzink Roofing and Wall Cladding* also recommends preservative-treated softwood decking with a continuous layer of bituminous roofing felt or other non-porous lining on top in order to protect the metal from alkali, condensation and wood preservatives (Rheinzik GMBH 1988). However, for lead we have found that bituminous linings can melt and adhere to the lead. In addition, as a consequence of lead's chemistry, even small amounts of moisture trapped between lead and impervious underlays can cause large amounts of underside lead corrosion from refluxing water in the absence of air, particularly if there are also any acid contaminants, which have sometimes been claimed to be present in bitumen itself, though Hoffman & Maatsch (1970, 293) do not agree.

16. the reason for a smaller outlet than inlet in the UK is probably to reduce the negative pressure in the air gap, which could otherwise draw additional moist air through. Though BRE could not confirm this (C Sanders, BRE Scottish Laboratory, pers comm), recent North American studies support this practice.

17. Another referenced document (DTO 40.21) suggests that humid Channel and North Sea coastal regions of France require similar care.

18. A Canadian paper recommends four-way CDR ventilation with a counter-battened arrangement (Shaw and Brown 1982).

19. In 1737 it was ascertained that the roof of Salisbury Cathedral contained 2641 tons of timber (Gwilt 1982)

20. In a ventilated warm roof it is good practice to put a breather layer which is impermeable to liquid on top of the insulation, detailed to allow any condensation or water ingress from above to run out into the gutter.

21. Exceptions are constant distillation, as in accelerated testing and in some failed warm roofs, situations where the lead is thin or highly stressed and a small amount of corrosion has a disproportionate effect, and sometimes in nosings surrounded by splashlaps which are often rainwater-filled.

22. One contractor told us that in this inflationary age, he finds 'pound gaps', slightly wider gaps, about the thickness of today's pound coin (just over 3 mm) to be more suitable.

23. Computer modelling by BRE Scottish Laboratory suggests that if lead is laid over dry softwood in September, this will help to protect the underside from condensation for the entire first winter. However, the model does not allow for short-circuiting of moisture through gaps.

24. In general, and owing to natural buoyancy effects, air will tend to enter the church at low level and leave at high level. Water vapour, being lighter than air, also tends to rise to high level though the effect is often small because the air is mixed by convection currents. The consequence is that, on average, the dewpoint will tend to be higher at high level. However, when the church is heated either artificially or by sunshine, it will tend to be warmer there too. Differences in the dynamics of heat, air and moisture transfer will affect the corrosion patterns observed.

25. The running-out of condensate into the laps is also a characteristic pattern where Type II corrosion is involved. While one might expect condensate to accumulate more around the line of the lap between the two sheets, for the porous arched structures observed with Type II corrosion, water can move around under the outer surface of the corrosion product. Indeed, on the day this photograph was taken, the outer surface of the corrosion product above the gap appeared to be dry, but when scraped it exuded moisture and turned into more of a paste.

26. We have not actually found any sites with bituminous felt layers alone, but in RTL's tests they did not perform very well even in the absence of acetic acid because of the corrosive effects of any ingressed and trapped moisture. They also tended to stick themselves to the lead.

27. This is where oak boarding is most often found because it also serves as the ceiling.

28. Some recent evidence suggests that, as it ages, timber becomes more permeable to moisture and increases in equilibrium moisture content (Ridout 1995).

29. A pressure of 1 Pascal is approximately equal to 0.1 mm WG (Water Gauge) on a manometer.

30. A study of the statistics of lead roof failures is beginning and will examine some parish records more systematically.

31. It is also possible that the age of coal, which brought with it high ventilation rates, dry atmospheres in buildings (at least if they did not have too much gas lighting), depressurization of roofspaces and a high-sulphur local environment might have been a golden age as far as lead survival was concerned.

32. In real life, however, the situation is dynamic, changing with season, time of day and activity within the building. This applies particularly to the timber immediately in contact with

the lead, whose temperature swings greatly exceed those of the outside air owing to warming by the sun and radiant cooling to the clear night sky. At the same time moisture will come and go to the lead/wood interface both through the wood itself, around the edges of the planks, and via joints in the lead.

33 With the possible exception of very old lead with a high content (> 1%) of other metals, which is usually very difficult to work.

34 The louvres have fans behind them which are intended (and used) for exhaust ventilation of the roof void in hot sunny weather only, in order to reduce radiant heat gains into the rooms underneath. However, since the fans do not have shutters, the louvres are always available for natural ventilation too, and the dominant air movement path is of warm air rising from within the building under natural buoyancy.

BIBLIOGRAPHY

Billiton Zink BV, *Advice TZ5: Structures with rear ventilation*, (Billiton Zink BV: Budel, Netherlands, 1992).

British Standards Institution, *British Standard 5250: 1989 BS Code of Practice for Control of Condensation in Buildings*, (British Standards Institution: London, 1989).

W. T. Bordass, 'The effects – for good and ill – of building services and their controls', in *Building Pathology 89 Conference Proceedings*, (Hutton + Rostron: Guildford, 1990), 39–68.

W. T. Bordass, J. Charles and D. M. Farrell, 'Corrosion and decay in sheet metal roofing', in *Building Pathology 90*, (Hutton + Rostron: Guildford, 1991), 64–9.

W. T. Bordass, G. Dicken, and D. M. Farrell, 'Corrosion control: sheet metal roofs', in *Architects' Journal*, 29 November 1989, 71–5.

William Bordass Associates, 'Moisture in the roof voids', in *St Cross Church: Direct Gas-Fired Heating. Monitoring Report 1985–86 Heating Season*, (William Bordass Associates: London, 1986), 40–49.

F. L. Brady, *The prevention of corrosion of lead in buildings*, DSIR Building Research Bulletin 6 (HMSO: London, 1929).

Building Research Establishment, BRE Digest 270, *Condensation in insulated domestic roofs*, (Building Research Establishment: Garston, 1983).

–, BRE Digest 301, *Corrosion of metals by wood*, (Building Research Establishment: Garston, 1985).

–, BRE Digest 310, *Flat roof design: the technical options*, (Building Research Establishment: Garston, 1986).

–, BRE Digest 324, *Flat roof design: thermal insulation*, (Building Research Establishment: Garston, 1987).

–, *Thermal Insulation: Avoiding Risks*, (HMSO: London, 1994).

Centre Scientifique et Technique du Bâtiment, Document technique unifié DTO No 40.44, *Couvertures par éléments métalliques en feuilles et longues feuilles en acier inoxydable étamé-plombé*, (Centre Scientifique et Technique du Bâtiment: Paris, 1991).

S. E. Clarke and E. E. Longhurst, 'The corrosion of metals by acid vapours from wood', in *Journal of Applied Chemistry* 11, (1961), 425.

P. Cleary and M. Sherman, 'Seasonal storage of moisture in roof sheathing', in *Air Infiltration and Ventilation Centre Technical Note 20*, 2 (July 1987), 2.1–2.20.

F. Coote, *Lead roofs - repair or renewal?*, (Lead Sheet Association: Tunbridge Wells, October 1994).

Department of Industry, *Guides to Practice in Corrosion Control: 2: Corrosion of metals by wood*, (Central Office of Information/HMSO: London, 1979).

G. Dicken and D. M. Farrell, *Monitoring of underside lead corrosion, Phase II final report*, (CAPCIS Ltd: Manchester, August 1990).

Ecclesiastical Architects' and Surveyors' Association, *Corrosion of Lead Roofing: Interim Report of a Sub-committee* (Ecclesiastical Architects' and Surveyors' Association: London, 1986a).

–, *Lead Corrosion Report: Draft Addendum Sheet*, (Ecclesiastical Architects' and Surveyors' Association: London, 1986b).

R. Edwards, *Atmospheric corrosion products of lead*, paper presented at the English Heritage Condensation Corrosion Forum (March 1994).

Energy Design Update, 'Building scientists turn construction code and common practice upside-down', in *Energy Design Update*, (June 1994), 3.

Eurocom, *The Eurocom Handbook: Terned Stainless Steel Roofing and Cladding*, (Eurocom Enterprise Ltd : Ascot, 1993).

D. M. Farrell, and G. Dicken, 'A review of condensation corrosion', in *Proceedings of UK Corrosion 90*, (1990), 2.71–2.81.

D. M. Farrell, A. S. Doughty, D. G. John and W. T. Bordass, *A review of corrosion of metal roofs*, (CAPCIS Ltd: Manchester, 1992).

Y. E. Forgues, *The ventilation of insulated roofs*, (Division of Building Research, National Research Council Canada: Ottawa, October 1985).

C. M. Grzywacz, and N. H. Tennent, 'Pollution monitoring in storage and display cabinets: Carbonyl pollutant levels in relation to artifact deterioration', in *Preprints of the Ottawa Congress: Preventive conservation practice, theory and research*, (International Institute for Conservation: London 1994), 164–70.

J. Gwilt, *The Encyclopedia of Architecture*, Book 1, (Longmans: London, 1867; Crown: New York, 1982).

R. H. Hill, *The condensation corrosion of metal roofing materials*, Lead Industries Group Ltd Research Laboratories Report MM/7/82 (September 1982).

W. Hoffman, and J. Maatsch, 'Lead as a corrosion-resistant material, in *Lead and lead alloys* (Springer-Verlag: Berlin, 1970), 268–320.

International Energy Agency, *IEA Annex 19 A guidebook for insulated, low-slope roof systems*, (International Energy Agency: Coventry, February 1994).

R. Jones, 'Modelling water vapour conditions in buildings', in *Building Services Engineering Research and Technology*, 14 (1993), 99–106.

Lead Contractors' Association, 'Condensation Can Bring Corrosion', in *Lead Contractors' Association Members Newsletter* 9 (16 October 1986).

Lead Development Association, 'Condensation - a problem ..... even for lead', in *Lead Development Association Bulletin* 5 (1988).

Lead Sheet Association, *The lead sheet manual*, 3 (Lead Sheet Association: Tunbridge Wells, 1993a), 91–3.

–, *Update 2: Underside Corrosion*, (Lead Sheet Association: Tunbridge Wells, 1993b).

–, *Control of lead at work*, (Lead Sheet Association: Tunbridge Wells, 1993c).

R. Lowe, K. G. Davies, D. M. Farrell, W. T. Bordass and R. Edwards, *Donnington Castle: Forensic examination of a lead roof*, (Rowan Technologies Ltd, unpublished 1994).

F. Loxsom, *Meteorological data for the passive solar programme*, (Research in Building Group, Polytechnic of Central London: London, 1986).

R. C. McLean and G. H. Galbraith, 'Interstitial condensation: applicability of conventional vapour permeability values', in *Building Services Engineering Research and Technology* 9, (1988), 29–34.

R. Murdoch, 'Avoiding corrosion under metal roofs', in *Building Technology and Management*, (October/November 1987), 4.

W. A. Oddy, 'The corrosion of metals on display', in *Preprints of the Stockholm Congress: Conservation in Archaeology and the*

Applied Arts, (International Institute for Conservation: London, 1975), 235–7.

T. A. Oxley and E. G. Gobert, *Dampness in Buildings*, (Butterworth-Heinemann: London, 1994).

M. Pourbaix, *Atlas of Equilibrium Diagrams*, (Pergamon Press: London, 1966).

M. Pourbaix, N. de Zoubov, C. Vanleugenhage and P. Van Rysselberghe, 'Lead', in M. Pourbaix 1966, 485–92.

B. Ridout, 'Larva Louts', in *Building*, (31 March 1995), 52–3.

Rheinzik GMBH, *Architecture with Rheinzink Roofing and Wall Cladding*, (Rheinzik GMBH: Datteln, Germany, 1988).

Rowan Technologies Ltd, *Investigation of underside corrosion of lead roofs*, Progress Reports 1–6 (Rowan Technologies Ltd: Manchester, 1993–5).

C. Y. Shaw and W. C. Brown, Canadian Building Research Note 192 *Effect of a gas furnace chimney on air leakage characteristics of a two-storey detached house*, (Division of Building Research, National Research Council, Canada: Ottawa, July 1982).

A. Simpson, G. Castles and D. E. O'Connor, 'Condensation, heat transfer and ventilation processes in flat timber-frame cold roofs', in *Building Services Engineering Research and Technology* 13, (1992), 133–45.

J. Stillman and J. Eastwick-Field, *The theory and practice of joinery*, (Architectural Press: London, 1966)

N. H. Tennent, J. Tate and L. Cannon, 'The corrosion of lead artifacts in wooden storage cabinets', in *SSCR Journal* 4, (1993), 8–12.

G. G. Tranter, ' Patination of lead: an infrared spectroscopic study', in *British Corrosion Journal* 4, (1976), 11.

W. H. J. Vernon, 'Second Experimental Report to the atmospheric corrosion research committee (British Non-Ferrous Metals Research Association)', in *Transactions of the Faraday Society*, 23 (1927), 113–204.

R. Watson, *Chemical Essays* 3, (1787), 366.

G. Werner, 'Corrosion of metal caused by wood in closed spaces', in J. Black (ed) *Recent advances in the conservation and analysis of artefacts,* (Institute of Archaeology Summer School, University of London: London, 1987).

AUTHOR BIOGRAPHY

Dr William Bordass has had a long practical interest in the design and performance of buildings. In 1970, following research in physical chemistry at Cambridge, he joined a multidisciplinary design practice where he developed an in-house building services engineering group and set up a specialist team concentrating on investigating environmental and energy performance. He was founder chairman of the London Energy Group and Professor of Building at the Bartlett School of Architecture and Planning at University College London (1987–8). In 1989 he was chairman of the Energy Conservation and Solar Centre in London, an educational charity.

In 1983 he set up William Bordass Associates which works in environmental control, energy efficiency, new technology, and physical and chemical deterioration in both new and old buildings. He is the author of the Council for the Care of Churches' book, *Heating your Church* and is currently monitoring the effects of heating and air conditioning systems in roof spaces. His work on lead roof corrosion for English Heritage, the National Trust and the Historic Royal Palaces Agency covers studies of materials, treatments, construction details and of roof space environments.

Sideflash in lightning protection: a report on tests with masonry

NORMAN L ALLEN
High Voltage Laboratory, Department of Electrical Engineering & Electronics, University of Manchester Institute of Science & Technology (UMIST), Sackville Street, Manchester, M60 1QD

Abstract

Risks of sideflash damage to buildings are ameliorated by either bonding all metalwork to the lightning protection system or by ensuring that metalwork is separated by a suitable distance (found by reference to British Standard 6651 [BSI 1992] after estimation of the likely maximum voltage). Sideflash distances through walls have been calculated as if in air. However, it is reported here that the breakdown voltage in masonry, about 260 kV/m, is almost half that for air and this has consequences for separation distances, depending upon the design of the lightning protection system. The results will aid risk assessments of existing lightning protection schemes and the design of new systems.

Key words

Lightning protection, sideflash, risk assessment tests

PREAMBLE

Research into sideflash in lightning protection of historic masonry buildings was commissioned by English Heritage because of concern from cathedral architects about the physical damage that can potentially occur during a lightning strike, whether or not the structure is protected by a lightning protection system, and the need to know how this risk might be minimized.

When a lightning conductor is struck at the top of a building, a very high voltage, perhaps of the order of a million volts, is developed at the point of contact. This voltage may be sufficient to break down the insulation afforded by the air or other intervening medium (for example, by masonry or timber) between the conductor and any earthed secondary conductors which may be inside or outside the building. Examples of such secondary conductors include smoke detector systems under the roof, electrical installations, water tanks and plumbing, tie rods in spires and drain pipes (Fig 1)

Although sideflash seldom occurs, it can be very destructive to electrical systems and may also cause serious structural damage. The physical forces generated during a lightning strike are formidable. Most of the energy released is in the form of heat, which can be disruptive to masonry and timber and which also vaporizes the moisture that is usually present with explosive effect. If large currents pass to earth through internal service cables or ducts, associated magnetic forces can cause serious distortion and tear them away from their fixtures, besides destroying the equipment they serve.

To ameliorate the risk of damage to buildings, the problem is countered by either bonding all metalwork to the lightning protection system by means of a copper conductor, or ensuring that metalwork is separated by a suitable distance. This distance is usually found by reference to British Standard Code of Practice 6651 *Protection of Structures against Lightning*, after estimation of the likely maximum voltage. Traditionally, sideflash distances through walls have been calculated as if in air.

Figure 1. Displacement and loss of solid ashlar stone blocks in masonry caused by sideflash effect at Easton Maudit church: the damage was caused below a spire-top air terminal and external conductor tape with lightning passing through the masonry between these features and an internal iron vane rod (Photograph: Tim Donlon/Church Conservation Ltd).

However, in this paper it is reported that the breakdown voltage in masonry, about 260 kV/m, is almost half that for air.[1] This means that the separation distances through walls must be increased by a factor of two depending upon the number of down conductors and earth electrodes in the lightning protection system which together determine the system voltage. Thick walls will obviously have more effect on sideflash calculations than thin walls. Where the necessary distance between metalwork cannot be obtained, bonding will be required.

Sideflash can, of course, also occur along the surface of a wall, as well as penetrating through it. This paper notes, however, that in this case the discharge takes place in the air and the performance is similar to that assumed in the Code of Practice.

Taking all the other risk assessment criteria into account as expressed in British Standard guidance, for most lightning protection installations the results of this research mean that more bonding of adjacent earthed metalwork should be considered, particularly in buildings having flammable or explosive contents, where the risk is more significant.

For existing installations that have performed successfully during lightning strikes, the implications may not be too serious although each system should be looked at on its merits during the next cyclical maintenance inspection. Naturally the impact on historic buildings protection will vary: in some circumstances, more earth bonding of metalwork will be required and its installation must be carefully handled to minimize its visual impact, but in other situations the distances between unprotected material, if correctly calculated, will be sufficient and no changes need be made.

The results of this research have been passed to the British Standards Institution for consideration and may well result, in due course, in a modification to Section 18 of BS 6651.

INTRODUCTION

When lightning strikes a lightning protection system on a structure, the current should pass safely to earth via the down conductors. However, because these conductors have a finite resistance to earth and a significant inductance between the point of striking and the ground, large transient voltages are set up in the system. These rise to a maximum at the point of striking, reducing towards the ground, but large enough at many points to spark through – sideflash – to nearby secondary conductors.

The Code of Practice, BS 6651, indicates the clearance required between the lightning conductor and an earthed conductor to prevent the occurrence of sideflash (Fig 2). It assumes however that the two elements are separated only by air; no account is taken of any building materials which may be present in the intervening space. Clause 18.2.2 of the Code states that 'the effect of any non-conducting masonry or other building material in the possible flashover path can be disregarded'.

This must occasion doubt about the advice given, since common sense suggests that the voltage that can be

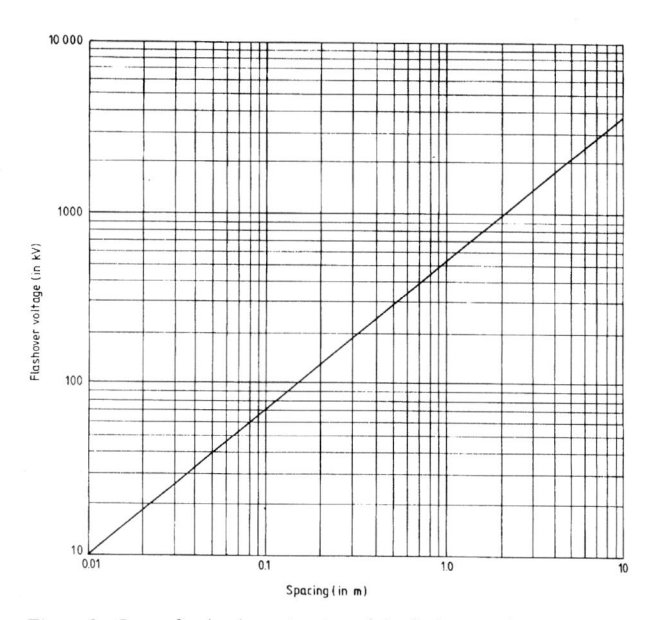

Figure 2. Curve for the determination of the flashover voltage in air as a function of spacing. Taken from Figure 28 in BS 6551 (1992) with kind permission of the British Standards Institution.

withstood by a given thickness of air will be different from that withstood when the same thickness of masonry is present. Further, it is well known in high-voltage technology that solid insulation which has a fault in the form of a crack is very vulnerable to sparking through the crack, and is usually 'weaker' against high voltages than the same thickness of air. Since cracks are common in masonry, laboratory experiments are needed to determine withstand voltages in simulated practical situations to provide realistic guidance for future practice.

It is worth making the point here, for future reference, that the slope of the line given in the Code is approximately 500 kV/m and that this is very close to the average electrical stress required to break down air under non-uniform field conditions.

Experiments were initiated in which configurations of common brick stacks were subjected to impulse voltages of the same time duration and order of magnitude as those experienced on buildings equipped with lightning conductors. This report sets out a summary of the results obtained.

TEST ARRANGEMENTS

Four experimental arrangements were designed and constructed to simulate conditions that may arise in practical protection systems.

Test Rig I

A stack of loose bricks was arranged as in Figure 3a. The brickwork was of simple, unbonded, single leaf form with a rectangular elevation. No mortar was used and the height was varied simply by adding or subtracting bricks. The bricks were stacked as close as possible to each other and care was taken to ensure that the central, vertical, through joint was as true as practicable. Even so, it is likely that the 'crack' which it formed was up to 2mm wide in

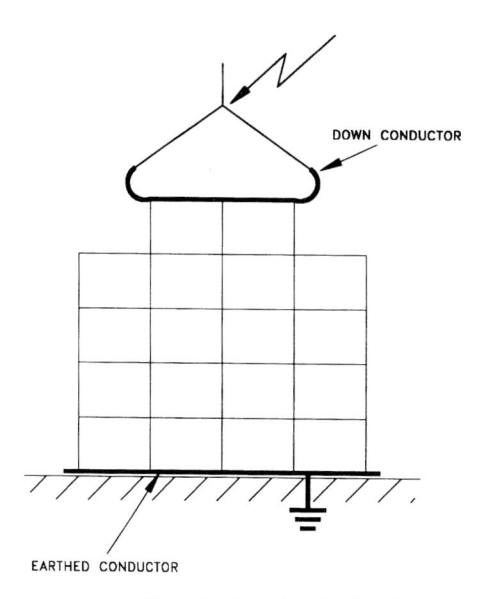

Figure 3a. Stack of loose bricks with strip electrode representing a down conductor.

places. The stack was built on an aluminium strip, which was earthed. Across the top of the stack was laid a strip of 6mm x 6mm aluminium to which impulse voltages were applied. The strip was bent away from the stack at each end to avoid spurious spark-overs. Thus, if this configuration is imagined to be turned through 90 degrees, it can simulate a down conductor placed opposite an earthed conductor of some kind on the other side of the wall, eg internal metal plumbing system or a metal staircase. Details of the voltage application are given below.

Test Rig II

An alternative situation was devised (Fig 3b) in which the down conductor was replaced by a more localized electrode, in the form of a rod, to represent other, arbitrary situations that may arise in a building, eg an electrical fitting or a tie rod. In other respects, the configuration was as in Test Rig I above.

Test Rig III

A vertical wall was constructed with mortar[2] and allowed to set for 28 days. It was of single brick thickness and the

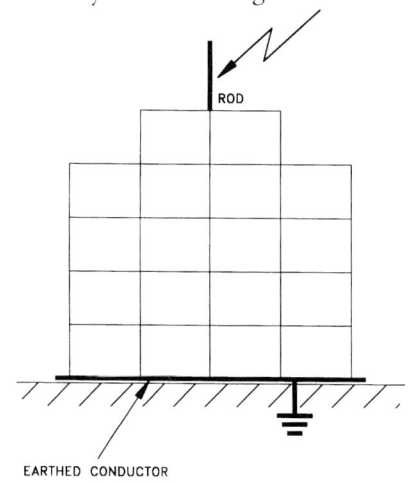

Figure 3b. Stack of bricks with rod electrode.

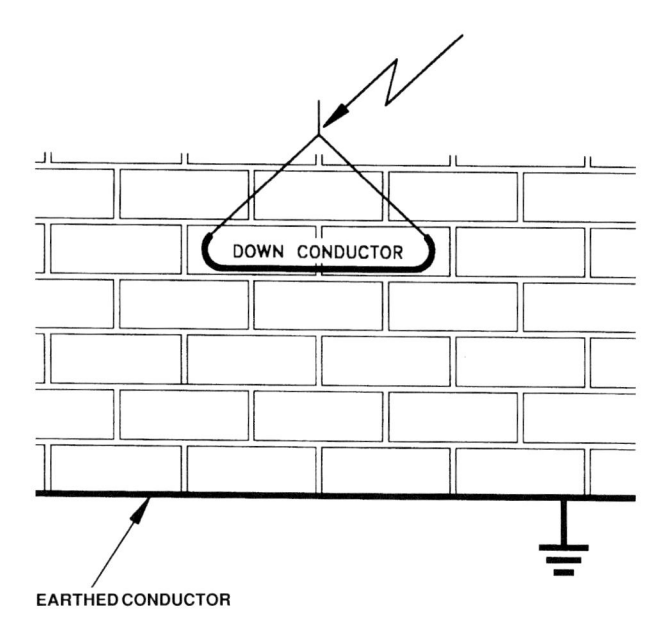

Figure 4. Electrode arrangement alongside a masonry wall of bricks and mortar.

dimensions were 1.55 metres (length), 870mm (height) and 102mm (width). The 6mm x 6mm strip conductor was lightly fastened[3] to the wall, at various heights above the base, and flashover tests carried out (Fig 4). This arrangement examines another form of sideflash, along the masonry surface from the moveable strip to an earthed metal strip fixed at the base. This simulates, for example, the sideflash from a down conductor to a neighbouring electricity cable or drainpipe on the same face.

Test Rig IV

The wall thickness was now increased to three bricks, approximately 325mm, using mortar, and allowed to dry this time for 38 days. The object was to test the conditions for spark-over through the wall which, since it was newly constructed, was free of cracks due to settlement. In this case, identical strip electrodes of aluminum 500mm x 50mm were lightly attached on opposite sides of the wall (Fig 5) at a height of 400mm above ground. The distance between the electrodes was thus 325mm. The impulse voltage was applied to one electrode; the other was earthed.

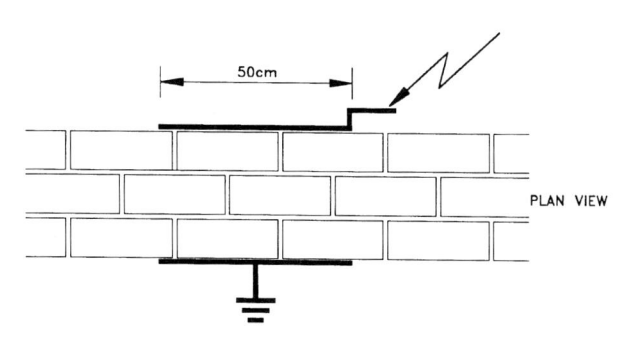

Figure 5. Test on sparkover through a brick and mortar wall.

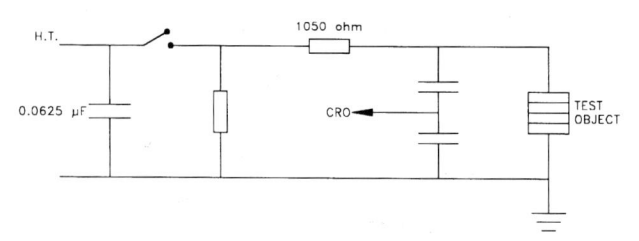

Figure 6. Equivalent circuit of impulse generator and test object.

Test Rig V

The arrangement of Test Rig IV was used to extend the tests to greater thicknesses of wall by staggering a pair of electrodes (now 100mm x 50mm) horizontally, so as to give distances between them of 500mm and 1.05m respectively. In addition a few tests were made with both the high voltage and earth electrodes separated by air thicknesses from the two faces of the wall to test the vulnerability of internal artefacts which may be set at some distance from the wall.

It should be noted here that in all cases where the electrodes were attached to the wall, they overlapped a horizontal mortar joint, since such joints were expected to present the greatest vulnerability to high voltage breakdown.

Pulsed voltages were applied to the high voltage electrodes (the conductor or rod) from an impulse generator capable of producing up to 800 kV. The voltage rose to its maximum value in about 1 microsecond and then decayed to about half the maximum in 50 microseconds, thus known as the 1/50 microsecond impulse. The equivalent circuit is shown in Figure 6. All recorded values of spark-over voltage were adjusted to the standard atmospheric density condition at 1013 mbar at a temperature of 20°C, but no correction for humidity variation was made. Most lightning flashes were negative, but some were positive. Both polarities have been used here, but emphasis has been given to positive, since it is more dangerous (ie flashes over at lower voltages) than negative. In all these tests, spark-over voltages were measured

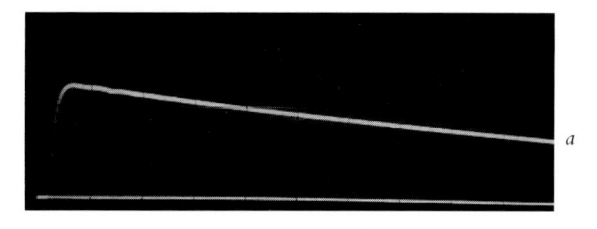

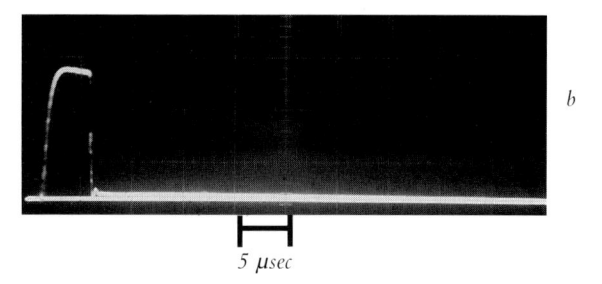

Figure 7. Oscillograms of lightning impulse voltage applied to the stack of bricks; (a) crest voltage = 160 kV: no spark-over; (b) crest voltage = 186 kV: spark-over at ~ 5 microseconds. Scale of each grid = 5 microseconds.

by the 'up and down' method using a total of 40 shots to obtain each mean value.

An example of an impulse voltage waveform applied to the test object, where no spark-over occurred, is shown in Figure 7a. Where a spark-over occurred, followed by a full power arc, the voltage collapse was as in Figure 7b: this example was obtained using the stack of bricks described in Test Rig I above.

RESULTS

Crack in brickwork: Test Rigs I & II

All the results obtained with the stack of bricks are shown in Figure 8. Flashover voltages are plotted for the two configurations, 2a and 2b, and compared with the BS 6651 curve (see Fig 2). Three cases are shown: positive

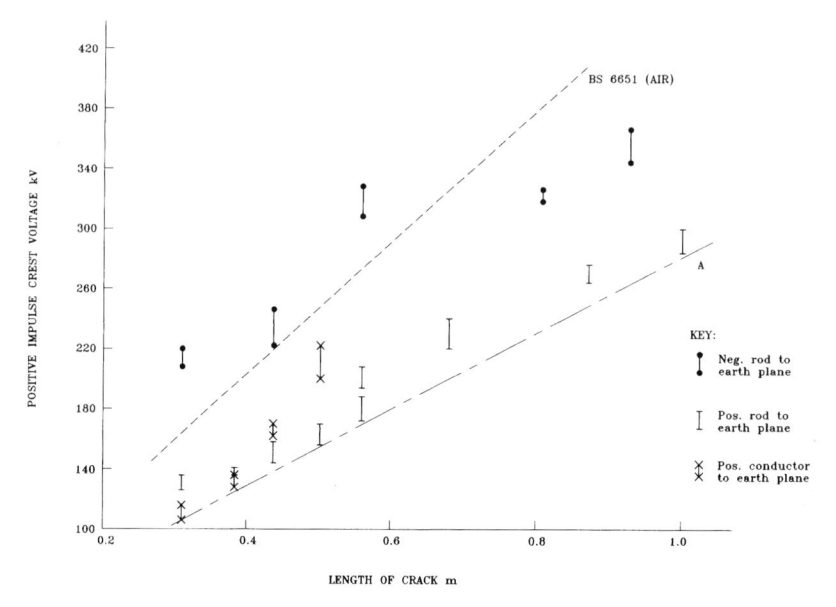

Figure 8. Flashover voltage versus length of crack.

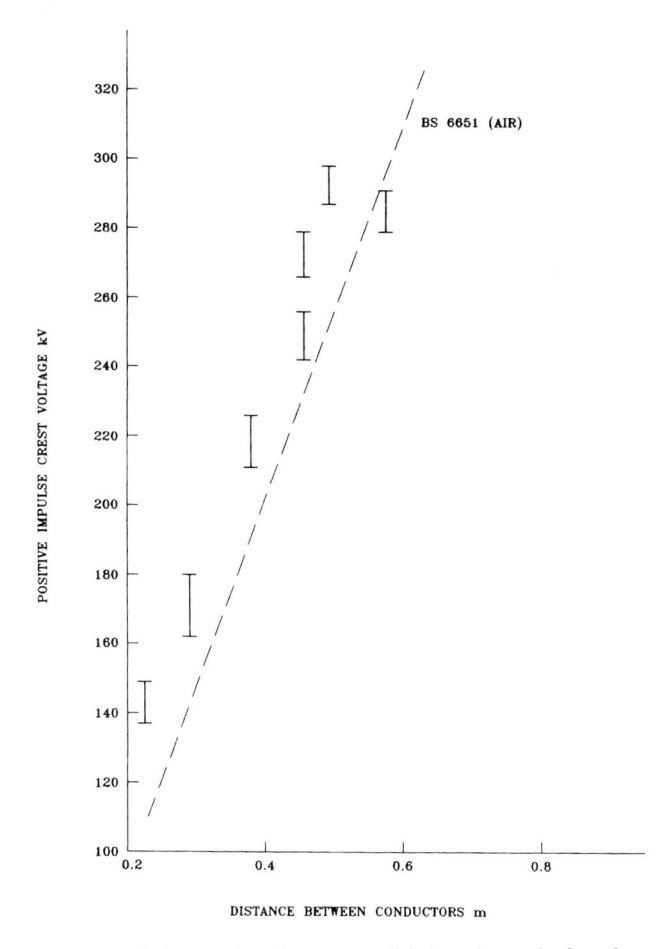

Figure 9. Flashover voltage between parallel electrodes on the face of brickwork.

strip conductor to earth and negative and positive rod to earth. The first case shows spark-over voltages significantly lower than those expected from the BS curve, but the last-mentioned, where the degree of non-uniformity of the field is greater, shows still lower values. In the case of the negatively-excited rod, spark-over voltages show no clear linearity with the length of crack. For smaller lengths they are greater than indicated by the BS curve, but the reverse is true at greater lengths. It is worth noting here that in air alone, spark-over voltages with a negative rod electrode would be expected to be about twice those predicted from the BS curve.

The line A is drawn to indicate the lowest voltages, for the range of clearances used, at which spark-overs occurred. The slope of this line corresponds to a mean stress, for spark-over, of 260 kV/m.

Flashover across face of brickwork: Test Rig III

A plot of flashover voltage as a function of distance between electrodes for the arrangement of Figure 4 is shown in Figure 9. Flashover voltages are very close to those indicated by the BS curve, which is therefore shown to be applicable, not only between electrodes in air alone, but also between electrodes in air with an intervening surface – rather than thickness – of brick and mortar.

It should be recorded here that the paths of the sparks along the brickwork face was sometimes variable. It was observed that in some cases, the spark would typically pass through the penultimate mortar joint before reaching the earthed electrode, and then pass to this electrode via the wood base upon which the wall was mounted. These sparks were counted with all others in assembling the data of Figure 9 and would have contributed to the spread of individual values noted there.

Spark-over through solid brick wall: Test Rig IV

The nature of the spark-over in this case was different from those in the two sets of configurations described above. Preliminary tests with the simple wall of single brick thickness showed that spark-over occurred readily at the lowest impulse voltages at which the generator could conveniently be operated, that is, of the order 50 kV.

The tests were repeated after the wall had been reconstituted as three thicknesses of brick. In this case, small sparks were observed at both the high voltage and the earth electrodes with a crest voltage of about 80 kV and above. The sparks did not, however have the appearance of full arcs, such as were experienced when the cracks were investigated, nor did the oscillograms show the same clear collapse of voltage. The experiments were therefore continued up to the largest crest voltage used, namely 330 kV.

Voltage oscillograms are presented in Figure 10. Spark-over did not develop to a full arc, of very small impedance, or resistance. After an initial collapse of voltage, a discharge of relatively high impedance followed, showing a voltage which decreased gradually during the length of the impulse. Comparison can be made with the oscillograms of Figure 7.

The impedance of the spark discharge can be estimated from the form of the voltage illustrated in Figure 10a. This oscillogram allows 'scaling up' of the measured voltage to the voltage actually appearing across the specimen under test. Also, knowledge of the impulse generator circuit (Fig 6) permits an estimate of current to be made. The resistance of the spark was found to be in the order of 500 ohms. This is very high for a spark; in the case of a fully developed spark through the crack, the resistance would have been a fraction of an ohm.

High resistance sparks, as these are known in the study of insulating materials, are characteristic of long breakdown paths which may be caused by faults in the specimen. The discharges are highly resistive due to loss of energy to adjacent solid surfaces as they progress through narrow interstices or channels. It is concluded, therefore, that in the case of the brick wall being tested the partial spark-overs occurred most probably at the interfaces between the mortar and the bricks, where trapped air allowed the discharge to follow tortuous channels.

It will be observed that this behaviour persisted up to the highest voltage used, namely 330 kV. It is assumed that where a voltage was sufficiently large to inject greater than a critical energy into the discharge channel, a full transition to the arc would occur. Some evidence that such transitions are likely is shown from the oscillograms

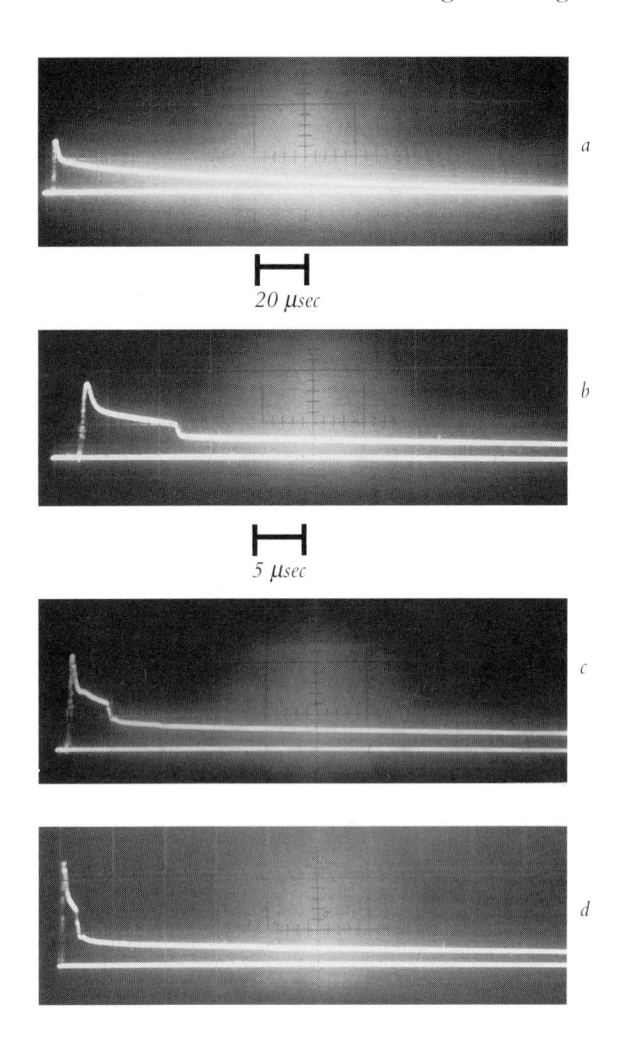

Figure 10. Oscillograms of lightning impulse voltages applied between opposite faces of a three brick thickness wall: crest voltages (a) = 162 kV, (b) = 196 kV, (c) = 296 kV and (d) = 330 kV.

at higher voltage levels, where a second transition to a lower impedance discharge was shown. This occurred earlier as voltage increased (Fig 10).

To summarize: the experiments showed that low-current spark-over through a wall of three bricks' thickness could occur at relatively low voltages, and that the observed sparks passed through at the brick mortar interface. With the dimensions of wall used in the tests work at higher voltages than about 350 kv was impracticable, but indications were that at sufficiently high voltages and energies, spark-overs leading to full power arcs might be expected.

OTHER TESTS WITH THE WALL

Test V

With the 100mm x 50mm staggered electrodes fixed to give a diagonal sparking distance through the wall of 500mm the minimum voltage at which spark-over occurred was in the region of 80 kV, at a very similar level to that for the results given for Test Rig IV, where the distance was 325mm. The electrodes were then moved to give a diagonal distance of 1.05m. The threshold voltage was again found to be of the order of 80 kV. Thus, the mean stress required for spark-over was very low, of the order of 100 kV/m, or less.

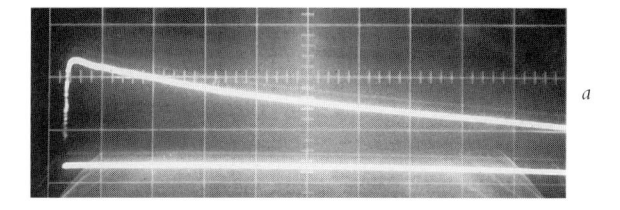

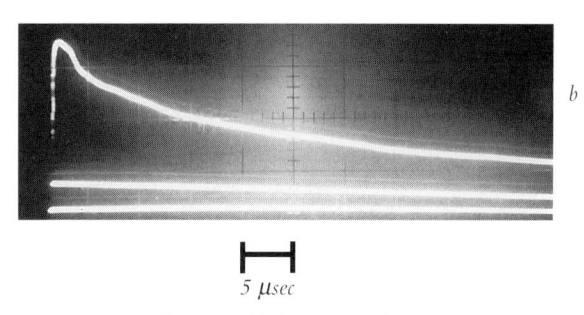

Figure 11. Oscillograms of lightning impulse voltages applied across a diagonal thickness of wall of 1.05m: crest voltages: (a) = 213 kV, (b) = 346 kV. Scale of grids (b), (c) and (d) = 5 microseconds.

Tests VI & VII

A series of higher voltage levels was then taken and oscillograms recorded to show the behaviour of the spark discharge. Two examples are given in Figure 11, at impulse crest voltages of 213 kV and 346 kV. At the former voltage there was no sudden collapse of voltage, but comparison with the applied impulse voltage shown in Figure 6a shows that the profile was perturbed by the formation of a limited, high impedance discharge and that at 346 kV a larger collapse of voltage, resulting from a higher-current spark-over, occurred. Thus the increased path length, compared with that used in Test Rig IV, resulted in a higher impedance discharge over the voltage ranges studied.

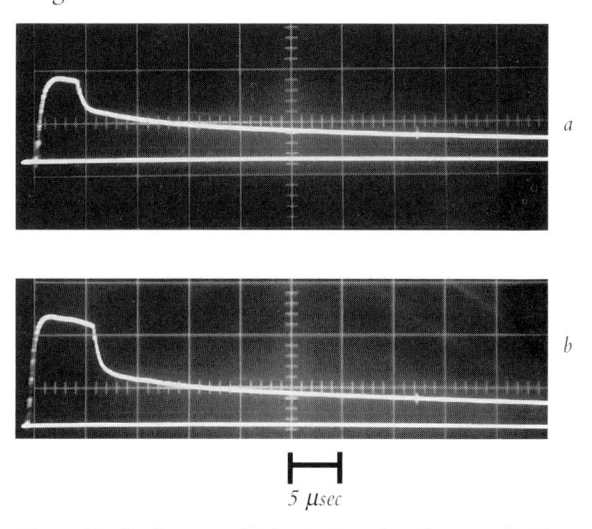

Figure 12. Oscillograms of lightning impulse voltages applied between a rod fixed at a distance from the three brick thick wall and a 100 mm x 50 mm electrode fixed to the opposite side: crest voltages at (a) 100mm = 114 kV (b) 200mm = 156 kV. Scale of each grid = 5 microseconds.

Tests VIII & IX

Tests were made in which the impulse voltage was applied to a rod, the end of which was fixed respectively at 100mm and 200mm from the wall. A 100mm x 50mm electrode was fixed, at the same level, to the opposite side of the wall. Spark-over occurred at mean voltages of 116 kV and 146 kV respectively, but, again, only partial collapse of voltages occurred. Sample oscillograms are shown in Figure 12.

Test X

When a 200mm air + 325mm wall + 200mm air arrangement was tried, the mean spark-over voltage was 305 kV, with partial collapse of voltage only. This again demonstrated that under none of the conditions used here did a power arc of low impedance follow the breakdown of the wall.

Test XI

Breakdown over the face of the wall was tried after it had been wetted with water. No significant difference was found between the spark-over voltages measured in this case and the dry tests (see Figure 9). The result is, perhaps, not surprising since the values obtained were characteristic of air: ie. the spark-over occurred in air and was not significantly affected by the proximity of the surface.

SPARK-OVER IN WOOD

A few tests were carried out with softwood joists, approximately 100mm x 75mm in section. First, a 500mm length was taken and impulse voltages applied along the grain using a metal sheet as earth electrode and the rod as positive high voltage electrode. The test was repeated with five beams stacked on their sides, so that the insulating qualities across the grain (total wood thickness 375mm, in this case) could be compared.

Along the grain, the spark-over voltage was 252 kV. Spark-over was observed to occur mainly in the air outside the wood, but towards the bottom, at the 'earthy' end, severe splintering occurred, indicating some penetration by the discharge. Across the grain, the mean spark-over voltage was 216 kV; sparks passed through the air and no splintering occurred.

It will be observed that in both configurations, the mean stress for spark-over was close to the value of 500 kV/m, which is characteristic of air breakdown (BS 1992, Fig 28). Tests with wood were not pursued further, owing to perceived difficulties of ensuring reproducible and meaningful conditions. Moreover, in practice, timber is normally embedded in other building materials, so that tests with isolated specimens would be unlikely to apply to the practical case.

CONCLUSIONS

The stress required for spark-over across a cemented brick surface under a positive lightning impulse voltage is 500 kV/m. This is approximately the same as that required for spark-over in air and as that given by Figure 28 of BS 6651 (*ibid*).

The stress required for spark-over through a crack in brickwork is approximately half that required for spark-over in air, under positive lightning impulse voltage, and approximately half that given by Figure 28 of BS 6651 (*ibid*).

Where negative impulse voltages are employed, the stress required for spark-over through a crack is greater than that under positive impulses, using the same electrode configuration.

The insulating qualities[4] of mortar joints in new brickwork, against high voltages, are poor. Sparks of relatively high impedance, with currents in the order of 100 amperes, are able to pass through mortar/brick interfaces, causing significant discharges at the electrode terminations when the applied stress is as low as approximately 100 kV/m. These discharges are unlikely to cause appreciable damage to the structure, but they could be dangerous where the interior of a building at that point contains flammable or explosive vapours.

Tests made with isolated pieces of softwood show stresses required for spark-over of 500 kV/m.

ENDNOTES

1 Units of measurement: as the leading consultants for most UK conservation projects are from the architectural and building surveying professions, and are used to the CI/SfB system of metric units, all measurements given in this paper for electrical field strength are set out as 260 kV/m (BS 1993) rather than the conventional format used by electrical engineers of 260 kVm^{-1}.

2 The mortar employed in the laboratory workshop was of a premixed, bagged type of cement and sand (1:3 by volume). Although this modern mix is not typical of, nor recommended for, use on historic buildings, it was not thought to be an important difference in the performance of the experiment. The discharges were probably in the interface between the mortar and the masonry, where air was trapped at the rough brick or stone surface. The constitution of the mortar would therefore not make much difference to the results.

3 The fastening process involved the use of adhesive tape. This adequately simulated the real-life situation in which conductors, though firmly stapled, are in light contact with the wall.

4 Electrical engineers often refer to the insulating qualities of materials in terms of 'strength'. The term is avoided here to limit confusion.

BIBLIOGRAPHY

British Standards Institution 5555, *Specification for SI units and recommendations for the use of their multiples and of certain other units* (London: British Standards Institution, 1993).

British Standards Institution Code of Practice 6651 *Protection of Structures against Lightning*, (London: British Standards Institution, 1992).

ACKNOWLEDGEMENT

The writer thanks D A Greaves for his help in arranging the experiments. Consultations on architectural matters took place with Harry Fairhurst FRIBA, formerly Surveyor of the Fabric of Manchester Cathedral. Diagrams were redrawn for publication by Jim Child of English Heritage.

The photograph of sideflash damage at Easton Maudit church was taken by Tim Donlon MBIOB of Church Conservation Ltd of Nottingham with the kind permission of J Barker of Gotch, Pearson Architects of Northampton on behalf of the PCC, and is reproduced here with grateful acknowledgement following publication in the *Building Conservation Directory 1997*, Cathedral Communications Ltd, Tisbury.

Extracts from BS 6651 (1992) are reproduced here with the permission of the British Standards Institution. Complete editions of the British Standards can be obtained by post from BSI Customer Services, 389 Chiswick High Road, London W4 4AL; tel: 0181 996 9000; fax: 0181 996 7001.

AUTHOR BIOGRAPHY

Dr **Norman L Allen** is Visiting Senior Lecturer in the Department of Electrical Engineering and Electronics at the University of Manchester Institute of Science and Technology (UMIST). A graduate of Birmingham University, he held research fellowships at the National Research Council in Ottawa and at the Massachusetts Institute of Technology, before returning to England to work with Associated Electrical Industries Ltd. After seven years in industry, he taught and researched in high voltage engineering for many years at Leeds University before moving, two years ago, to UMIST, where he is continuing his researches.

Part II

Development

The application of cathodic protection to historic buildings

Buried metal cramp conservation in the Inigo Jones Gateway, Chiswick House grounds, London

KEITH BLACKNEY
37 Goss's Lane, Cheshunt, Herts EN7 5EG
BILL MARTIN
English Heritage, Architectural Conservation Team, 23 Savile Row, London W1X 1AB

Abstract

The Inigo Jones gateway which stands in the parkland surrounding Chiswick House, an English Heritage property in London, had been the subject of at least three different restoration attempts. By 1990 it was considered to be in such poor condition that action needed to be taken to address the main causes and symptoms of deterioration, something which the previous interventions had signally failed to achieve. These interventions had only served to compound the decay mechanisms. One of the most serious threats to the stability of the structure, in common with so many historic buildings, was the corrosion of ferrous cramps embedded in the masonry. The accepted practice of cutting out or dismantling to remove such fixings was not considered appropriate and instead a method of non-destructive electrochemical rehabilitation, impressed current cathodic protection (ICCP), was explored and developed for use on the structure.

Key words

Ferrous cramps, cathodic protection, impulse radar, risk analysis

CHISWICK HOUSE AND THE INIGO JONES GATEWAY

Chiswick House (Figure 1) is situated within parkland at Chiswick, now part of metropolitan London. The house is in the care of English Heritage and the grounds are maintained by the London Borough of Hounslow. The villa, built 1726–9, was designed by Richard Boyle (1694–1753), third Earl of Burlington and fourth Earl of Cork. Burlington's aristocratic background and rounded education had groomed him to be a patron of the arts, dubbed the 'Apollo of the Arts' by Horace Walpole and 'the modern Vitruvius' by Daniel Defoe.

Burlington was the most prominent gentleman architect of his day. In 1714 he undertook the Grand Tour, one of the first of generations of English noblemen so to do, returning from Italy the following year to take charge of the family estates. Initially following the established Baroque fashion, the family had retained James Gibbs as their architect. It was not until 1717 and the replacement of Gibbs by Colen Campbell, who had just completed the first of his series of volumes *Vitruvius Britannicus*, containing drawings of Roman-inspired buildings he described as 'regular buildings', that Burlington began to commission buildings to his own designs. Following Campbell's efforts in reviving the pioneering work of the architect Inigo Jones of some 90 years before, Burlington was drawn to the classically-inspired architecture of sixteenth-century Italy, exemplified by the stone mason turned architect Andrea Palladio. Travelling again in Europe in 1719, Burlington returned with the multi-talented artist and designer William Kent who replaced

Figure 1. Chiswick House, London (English Heritage). See also Colour Plate 11.

Campbell and became Burlington's principal collaborator. The stone gateway known as the Inigo Jones gateway, less the rendered flanking walls, was designed by Jones and is believed to be heavily influenced by a gateway at the Villa Lante, the work of the Italian architect Vignola. Made for the Earl of Middlesex and erected along the Kings Road at Beaufort House, Chelsea in 1621, the gateway was given to Lord Burlington by Sir Hans Sloane in 1738, when the flanking walls were most probably added. The gateway is described as being in the Doric order and while this is true a closer definition, advised by research after the time of Jones, would be Roman Doric. Of it Pope wrote:

> I was brought from Chelsea last year
> Batter'd with wind and weather
> Inigo Jones put me together
> Sir Hans Sloane let me alone
> Burlington brought me hither[1]

The gateway, one of only seven surviving fully authenticated buildings by Jones, stands in the grounds at Chiswick as a lasting tribute from Burlington to one of his main sources of inspiration. The influence of both architects is also recognised in Burlington's commissioning in 1744 from John Michael Rysbrack, one of the greatest sculptors of the day, of the two statues of Jones and Palladio in Portland stone which now stand in front of Chiswick House.

The Boyle family had owned the estate at Chiswick for some time before the construction of the current building. A Jacobean house stood on the site acting as a country retreat from the bustle of London. The villa was intended more as a stylistic exercise than a major house. However, in 1733 Burlington moved his main south of England residence away from London to Chiswick. Consequently, more room was needed to house the family and attendants and the villa was joined to the old house by a link corridor. When Burlington died in 1753 Chiswick passed to the Cavendish family, as his daughter Charlotte had married William Cavendish, Marquess of Hartington, who became the fourth Duke of Devonshire in 1755. In 1788 the fifth Duke had the Jacobean house demolished and commissioned the architect John White to design flanking wings for the villa. From the 1840s the house was let to a succession of tenants, the last using the building as a mental hospital from 1892 until 1929. The house was then purchased by the Middlesex County Council and leased to the Borough of Brentford and Chiswick. In 1948 the building was bought by the government and in 1950 a programme of restoration was begun, the main feature of this being the demolition of the Georgian wings.

Condition in 1990

In 1990 a survey was commissioned by English Heritage to assess the condition of the gateway and to provide suggestions for remedial treatment. The survey identified three main forms of decay affecting the Bath stone structure.

There was a build-up of dirt and other deposited material on all surfaces but most noticeably on the vermiculated ashlar and on the undersides of the mouldings. This build-up had resulted in the formation of skins of calcium sulphate, the product of the reaction of the calcium carbonate of the Bath stone with the sulphur dioxide and trioxides in the urban atmosphere. These skins are discontinuous and characteristically black in appearance by virtue of the hydrocarbon particulate matter bound onto the stone surface by the gypsum layers. They eventually craze and blister and leave exfoliated craters exposing softer impoverished stone beneath due to the continuing action of soluble salt

Figure 2. The decayed surface of the Bath stone gateway showing inappropriate render repairs (English Heritage). See also Colour Plate 12.

Figure 3. A lens blown from the stone of the gateway as a result of cramp expansion (English Heritage).

crystallization cycles beneath the sulphate skin (Figure 2). Hard repair mortars had been used in areas of previously deteriorated stonework. Because of the impermeability of these repairs and the difference in density between them and the surrounding stone, the soluble salts present in the stone at the time of treatment, were either entrapped behind the patches, where they acted on the underlying stone to produce zones of deeply-seated erosion, or they acted on the stone surfaces at the edges of the repairs. These phenomena were recorded all over the arch but particularly noted on the south elevation where the cornice and the entablature were very heavily patched. There was also the problem of iron cramps rusting. Several instances of widening joints and cracking and jacking of surrounding stone and structural destabilization were recorded.

The overall picture of the gateway showed a series of piecemeal repairs in the past bringing it to such a state that they were contributing to decay. A more thorough conservation programme was thus required.

The remedial treatment recommendations contained in the survey were based upon a policy of minimum interference with the original fabric. These would include the removal of rusting iron cramps where this was deemed neccessary. Not long after the survey took place a section of masonry beneath the pediment on the west end of the gateway collapsed when the ferrous fixings in that area reached a critical level of corrosion.

Ferrous fixings in historic buildings

In the second of his *Ten Books on Architecture*, Vitruvius tells the reader that in building *opus reticulatum* he who wishes to avoid disaster should 'leave a cavity behind the facings, and on the inside build walls two feet thick, made of red dimension stone or burnt brick or lava in courses, and then bind them to the fronts by means of iron clamps and lead'. He goes on to say that such an arrangement 'will be strong enough to last forever'.[2]

This system of building was common until the middle of the last century when non-ferrous materials finally ousted iron as the main fixing material for construction methods. The change had of course been brought about by the long-awaited arrival of readily available materials that would not be as susceptible to the deleterious effects of moisture ingress in the same way iron had proved to be. Iron fixings, even if originally bedded in lead, are affected over the centuries by the ingress of dampness in weathered structures, and rusting occurs. The products of the corrosion can be many times greater in area than the parent iron and the powerful, localized forces imposed upon the masonry around the fixing can be sufficient to fracture the stone, blow large lenses from the face of the masonry (Figure 3) and in the worst cases destabilize the whole structure. The damage caused to historic buildings in this way costs hundreds of thousands of pounds per year to remedy.

As anyone who has looked in detail at repair and maintenance regimes for stone or brick buildings will be fully aware, the eventual rusting and expansion of buried masonry fixings can be the most serious problem affecting the integrity of both the structure and the historic material. This is often owing to the levels of intervention required to remove the rusting items and to replace them with other fixings. The removal process usually involves either the dismantling of the affected area of the structure to allow the deteriorated fixings to be removed and new fixings introduced or, if dismantling is either inappropriate or impossible, then the face of the building material in the area affected above and below the joint is cut out to a depth where it is possible to access the rear fixing point of the cramp or dowel. With the new fixing in place the face of the elevation is repaired with an indent piecing. These individual piecings will normally be placed 75–100 mm on the bed, depending upon the size of the void, and should normally be fixed with lime or other suitable mortar. All too often these are fixed in place with resin, either epoxy or polyester. Whichever one of these methods is employed it is certain that there will be tremendous disruption both in terms of collateral damage, the necessity of replacing disturbed stones, loss of archaeological evidence and long-term disruption of visual continuity.

Risk analysis of cramp removal

Following the collapse of the section of masonry The Architectural Conservation Team (ACT) of English Heritage was asked to inspect the gateway to comment on the proposals put forward as the result of the earlier survey and to suggest a reasonable programme structure for remedial treatment. At this time the structure was surrounded by scaffolding making it possible to examine each area of the gateway, particularly the arch, in great detail.

In general terms there was agreement with the diagnosis of the causes of deterioration as outlined in the survey report. There were, however, a number of areas concerning the level of intervention which it was felt were open to modification. These were the extent of cement repair removal and the removal of ferrous cramps, both of which were in most cases inexorably linked. A close examination of the surfaces confirmed that the levels of cement repair had gone even further than had been realised and the remedial intervention was reassessed.

The case for removal of cement repairs where they were associated with active and evident decay was accepted, but there was great concern that a policy of overall removal would not only create a project of a most unpredictable nature but that it would also be impossible to assess beforehand exactly what sort of surface would be left once the removal campaign had finished. In addition, the open-endedness of such a project would be liable to produce spiralling costs that would be difficult to accommodate in the yearly budget operated by English Heritage as a government-funded organisation. Given the above the normal practice for dealing with embedded iron cramps needed to be reviewed as this would have meant relatively large-scale dismantling with the added risk that even more extensive demolition would be required once the opening-up exercise had been initiated.

This pattern of events is common to many conservation repair projects worldwide. A programme initiated on the basis of minimum intervention can by almost organic development turn into a classic 'once and for all' repair programme with resultant costs in terms of loss of the structure's historical integrity, its archaeological importance and financial status. Such programmes are often the victims of insufficient risk evaluation, the need to establish a set of relative values for the object itself and the processes and conditions affecting its likely future and, from these, the need to present a strategy for conservation based on sustained maintenance in preference to large-scale intervention. To conserve structures through the implementation of such a strategy, it must seek to minimize the risks impacting on those structures in three main ways:

- eliminate or mitigate the causes of deterioration
- hinder weathering processes
- increase the resistance of the construction and materials.

Evaluation of the risks associated with the Inigo Jones gateway led ACT to produce a strategy involving the absolute minimum of physical interference combined with a facility for remote monitoring of the most sensitive of the deteriorating structure components, the cramps. Central to this evaluation was the recognition that the roof of the structure had been devoid of adequate protective covering for a long time, which had allowed more water into the core than would normally have been the case.

Such ingress had been instrumental in the continuing deterioration of the stonework and the rusting of the embedded iron fixings. Central to this thesis was the belief that the cramps could be successfully accessed and the corrosion passivated *in situ*. ACT were aware that a range of non-destructive electrochemical techniques had been developed for the rehabilitation of deteriorated steel in reinforced concrete and that civil engineers had increasingly turned to such techniques, often as the only means available to remedy decay that would otherwise lead to premature structural failure. The team recognised that one particular treatment, *impressed current cathodic protection* (ICCP) might, with development, be suitable to halt the corrosion of ferrous fixing components within historic masonry.

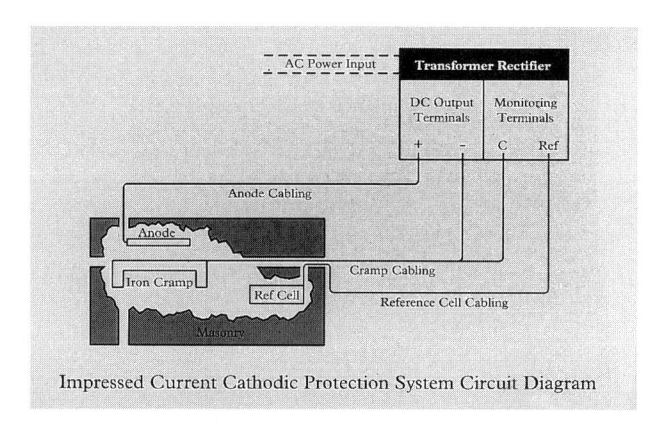

Figure 4. Impressed current cathodic system (ICCP) circuit diagram (English Heritage).

Cathodic protection

The first recorded work to purposely corrode a metal to protect another metal by making it a cathode, hence cathodic protection (CP), was carried out by Sir Humphrey Davy in the 1820s to protect the copper sheathing of ship's hulls. Davy's pioneering technique is still used today in both marine applications and for the preservation of buried underground structures such as oil pipelines. For metalwork exposed to the atmosphere metal-rich coatings are sometimes used, the coating acting as a barrier while the metallic content acts sacrificially at interruptions in the coating formed at the time of application or by later damage.

The big leap forward in cathodic protection technology has, since the 1950s, been ICCP. With ICCP a direct current (DC) generator, usually a transformer rectifier (TR), fed from an alternating current (AC) mains power supply, is series-connected between the anode and cathode, with the electrodes being respectively connected to the generator's positive and negative terminals. The generator can be adjusted to allow very precise control of current flow and drive voltage. The cathodic protection system works by converting the whole of the metal to be protected into a cathodic area and moving the anodic reaction to a more durable material, the current made at least equal to the corrosion current between the cathode and anode. The use of a power source has taken away the absolute requirement for anodes to be made of a more active metal to that requiring protection and as a consequence anodes are now made from a wide range of materials. The development of anode technology has resulted in allowing high current flows from small surface areas and a very long service life when compared to conventional sacrificial anodes. Figure 4 shows the basic elements of an ICCP system. One of the main attractions of ICCP for civil engineers is the system's ability to combat the effects of the free chlorides sometimes encountered within concrete. The chloride ion is highly aggressive to iron and will readily cause deep pitting of the metal's surface. The ionic current flows set up by ICCP will draw the negatively-charged chloride ions away from the steel reinforcing bars and towards the anode. A further advantage of using cathodic protection is that the cathodic reactions may gradually produce a more alkaline environment around the iron, further aiding its protection. ICCP is an active treatment that needs to be carefully controlled to maintain optimum performance. Reference electrodes are used to determine the condition of the metal and enable accurate setting of current and voltage, by supplying a known reference point against which the potential of the iron is measured. The unique aspect of the Inigo Jones project was that, until then, ICCP used in concrete or, rarely, stone was protecting continuous embedded metal. The Chiswick project would prove a testing ground for the location and connection of many, electrically isolated iron fixings, the situation found in most historic structures. The potential to develop the basically non-destructive nature of ICCP for conservation is enormous and the problems with the Inigo Jones gateway made the project an ideal testing ground to research, develop and apply a number of new and innovative treatments for the conservation of fragile historic masonry.

Survey

Whatever the method used to tackle the problem of the Inigo Jones cramps it was clear that the structure would first need to be surveyed to ascertain the overall condition of the masonry and the number, condition and exact position of its metal fixings. The criteria for selecting a survey method were that it must be non-damaging to the structure's fabric, logistically and technically viable for use on the scaffold around the gateway situated in a busy urban park and be sufficiently accurate to enable successive remedial treatments to be carried out accurately.

The method selected, pulsed radio echo sounding or impulse radar (IR), has been used for many years in geological survey, and more recently for the investigation of man-made structures where the technique has been used to measure material thickness and to locate voids and buried metals. These features indicated that not only would radar locate the metal fixings but that it would also give a good indication of their condition and effect on surrounding masonry. IR is non-destructive, relatively portable, does not require the cordoning-off of a large test area that would be necessary with some other non-destructive testing systems, for example radiography, and its survey data can be readily converted to produce very accurate working drawings. GB Geotechnics Ltd of Cambridge (GBG), a leading non-destructive testing (NDT) house, were consulted on the various NDT survey techniques and confirmed that IR would offer the accuracy and clarity needed to assess the structure both in its components parts and overall. GBG have pioneered the development of radar surveying, initially using the technique to detect subterranean faults beneath motorway carriageways and more recently to examine above-ground structures, for example high-rise accommodation blocks. As the name suggests impulse radar works by transmitting a pulse or plane of energy, approximately 50,000 times a second, into the structure, usually parallel to the direction of the survey. As each pulse travels forward through the structure it will meet changes in structure and these will reflect some of the pulse energy back to a receiver. Transducers of different frequency are used to emit the pulses, selected to be appropriate to the structure being surveyed and the anticipated targets: higher frequencies can resolve smaller targets while lower frequencies can achieve better penetration. The transducer is moved over the structure's surface to build up a cross-sectional analysis.

The Inigo Jones gateway was divided into horizontal survey lines along joints between ashlar and where possible through the centre of each stone. Level readings were along each horizontal line. However, this approach was modified at the cornice and pediment where the survey lines were drawn at angles. A high frequency 1000MHz transducer was used under a range of settings. The gateway was surveyed from both elevations and the information stored and taken to GBG's laboratories near Cambridge for analysis and digital re-mastering into working drawings. The survey revealed a range of fixings, cramps and dowels, and also provided information on construction elements of the gateway such as block depths, relationship of core to block formation and precise dimensional relationships of the structure overall. The information was given on working drawings with a key identifying type and orientation of cramps, non-ferrous fixings and areas of non-ferrous wire armature. Micro-cracks were identified in the masonry and the condition of cramps specified. Cramp locations were identified both pictorially and by coordinates from identified joints between ashlar, and cramp depths were also given. Some inconsistencies concerning the apparent presence of fixings in areas where, according to English Heritage experience, there was no reason for such a presence would be explored when practical location trials were carried out.

ICCP trial and testing of IR survey

English Heritage had been in discussion with Taywood Engineering Ltd (TEL), concerning the principles of ICP and the potential for adapting the technology for traditional masonry structures. At this time TEL were installing ICCP to prevent the further corrosion of a limestone-clad steel beam over the College of Sciences Colonnade now called Government Buildings, in Dublin.

English Heritage considered that in principle it would be possible to cathodically protect wrought iron, the main ferrous material for cramps and dowels, and that brick and limestone would carry sufficient moisture to enable the even distribution of protection current. It was necessary to conduct trials to prove the theory and the Inigo Jones project was selected as a representative, and available, example of the type of structure regularly found to be damaged by corrosion of ferrous fixings. Having located the cramps through IR survey the main challenge to overcome before the installation of ICCP to the gateway was to devise a methodology which allowed connections to the numerous electrically-discontinuous cramps and the planting of anodes and cables without further destabilizing the already fragile masonry. English Heritage commissioned TEL to carry out laboratory trials

Figure 5. Cross-sectioned model to demonstrate banana plug (English Heritage).

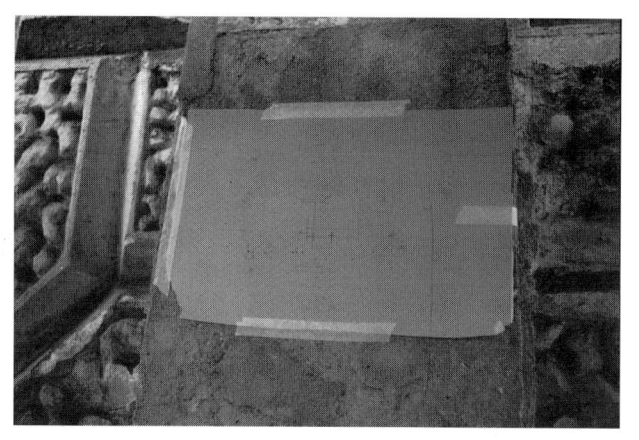

Figure 6. Cramp plotted onto card from cover meter readings (Taywood Engineering).

to establish methods to orientate short lengths of steel reinforcing, representing cramps, in concrete blocks, and to make cable connections to the cramp without resorting to the usual method of self-tapping screws each requiring excavation of approximately 50mm diameter. Detection of the steel was accomplished using a cover meter, a metal detector used to establish position and depth of cover of steel reinforcing in concrete location. This was successfully achieved in the laboratory trial. Meanwhile, a system of keyhole surgery was devised to make the cable-to-iron connection. The criteria for this system were:

- to be electrically and mechanically sound
- to connect a small diameter wire to an embedded cramp at 230mm depth using a working envelope of a 10mm diameter hole
- to be suitable for use (even if some further development were needed) during a large-scale ICP installation.

The keyhole system developed worked by drilling a 10mm hole through the masonry to the cramp. A steel sleeve was placed in the hole and a series of steel extension tools were passed through the sleeve to the cramp, which was spot-faced to present a flat surface, and then drilled with a centre drill followed by a 2.3mm diameter drill to a depth in excess of 8mm. Then, 0.8mm diameter polytetrafhoroethylene covered wires, coded black, were prepared by the addition of heat-shrunk polyethylene tubing. The wires were then crimped and soldered to a push-in 'banana' plug. The wires were fed inside a narrow tube until the base of the plug was held against the tube's end. The metal end of the plug was painted with a conductive silver-rich paint and the tube was passed through the masonry sleeve and the plug inserted into the hole in the cramp (Figure 5). Once inserted the tube and sleeve were withdrawn leaving the plug secure in the cramp and the wires protruding from the hole in the face of the masonry.

Site testing

To test the work carried out in the laboratory a joint EH/TEL pilot project was carried out to locate and cathodically protect two of the Inigo Jones gateway cramps. The initial challenge was to locate the cramps and to make a system-negative connection onto the cramps using the laboratory-developed application tooling. Using the GBG survey as a guide, TEL employed a cover meter for more accurate localised cramp location. Masonry experts from English Heritage drilled into the structure to access the cramps. The second part of the field exercise was to establish the feasibility of cathodically protecting the cramps by setting up a test CP system which would be monitored for two weeks. Two cramp locations were chosen from the GBG survey, one which seemed to offer a fairly straightforward target behind the flat surface of the pilaster on the west of the north elevation and the second a more demanding location underneath the east side of the architrave on the south elevation.

The first cramp was located with the cover meter; one addition to the method developed in the laboratory being a thin piece of card held against the stone surface allowing the meter head to travel more easily by helping to level out the surface. This method helped to establish the boundaries for the lowest reading regions which are set by a change of only 1mm (Figure 6). First indications that the IR survey would not be sufficient alone in locating the cramps came when both the cover meter survey and the opinion of English Heritage staff predicted correctly that the cramp in question was end-on to the pilaster face, whereas the IR survey had plotted it as laying parallel.

Figure 7. Special tool used to make the cramp connection (Taywood Engineering).

Figure 8. Trial installation on the gateway (Taywood Engineering).

The cramp was accessed and the connection process was successfully carried out (Figure 7). The second cramp, chosen to test the location methodology more rigorously, proved impossible to locate at all. Cover meter readings indicated proximity of metal but the apparent locations did not all concur with the plotting of the IR survey. A 5 mm hole was drilled at a point where 90 mm of cover was indicated, yet after drilling to a depth greater than 300 mm no contact had been made with the cramp. It seemed that in these types of location the cover meter would not be a sufficiently reliable indicator of position and it was therefore decided to drill in another area using just the data provided by the IR survey. The initial hole drilled on this basis also failed to make contact with a cramp. English Heritage's masonry expert drilled in from this hole at differing angles using his experience of masonry construction to try and predict where the cramp would be. This method also failed. Eventually a second cramp was located by moving to an area where the developed system was reckoned to have a greater chance of success. The location field trial had clearly demonstrated that even in reasonably straightforward areas no one method of cramp location would be sufficient, and that the combination of IR survey, cover meter and experience of masonry construction seemed to offer the best chance of success but even then there was an uncomfortably wide margin for error. It was decided that GBG would be asked to resurvey the structure to a tighter specification based on the experience gained from this element of the field trial. The second phase of the trial, the ICP installation, proved to be rather more successful. TEL installed a temporary system on 25th March 1992; and monitored the setup on three occasions: 28th March, 4th April and 9th April 1992, before the system was disconnected. The trial installation is shown in Figure 8, and its components were:

- two temporary platinized titanium rod anodes, the first placed on the surface of the gate, the second embedded in concrete mortar
- a constant current DC power supply
- system-negative connections of two types
- an embedded redox reference electrode (monitoring point 1).

Potentials at the stone surface at monitoring points 2 and 3 were recorded using a hand-held silver chloride half cell. The following parameters of the ICP system were monitored:

- the power supply voltage
- the applied cathodic protection current
- the current on potential of the cramps
- the instant off potential of the cramps.
- the potential of the cramps at various times after the current had been turned off.

Using initial results to define the cathodic polarization curve, the relationship between the application of ICP current and the instant off potential of the steel cramp, a level of current of 7 microamps was used for the trial. The instant off potential is recorded immediately after the ICP current is interrupted and is a measure of the true steel potential in its 'shifted state'. Potential decay is determined from the instant off potential and potentials recorded with the current off after a period of time. A power supply current of 7 microamps was selected as it produced a negative potential shift of the steel cramp of more than 200 mV from its base potential. These potential shifts recorded at each monitoring point during the first day all indicated that adequate protection was achieved.

Resurvey

In the autumn of 1993 GBG carried out a second phase of their original IR survey in an attempt to more precisely locate the cramps. On this occasion the locations would be marked on the stone surface with a removable marker.

The resurvey, employing a combination of pulsed radar and high-resolution metal detectors concentrated on those areas identified by the previous survey as being cramp locations. On this occasion a masonry expert from English Heritage was in attendance and it soon became apparent that some of the locations previously identified were not structural fixings but consisted of buried non-ferrous wire armatures and reinforcement for the extensive areas of repair render applied in one of the previous treatment programmes. Once again the importance of a masonry expert in the interpretation of specialist data was

FERROUS CRAMP LOCATION: SOUTH ELEVATION

CRAMP DESCRIPTION (JOINT / NUMBER)	DISTANCE OF CRAMP TO LHS OR RHS OF VERTICAL JOINTS (mm) ABOVE OR BELOW	APPROX. DEPTH OF COVER TO CRAMP (mm)
B/1 ?	RHS 100 BELOW	150
B/2 ?	RHS 182 BELOW	?
D/1 ?	LHS 390 BELOW	450
D/2 ?	LHS 130 BELOW	400
D/3 ?	LHS 130 BELOW	400
D/4 ?	RHS 200 BELOW	400
D/5 ?	RHS 310 BELOW	400
E/1 ?	RHS 123 BELOW	100
E/2 ?	RHS 445 BELOW	100
E/3 ?	RHS 1700 BELOW	70
E/4 ?	LHS 175 BELOW	80
F/1	RHS 20 BELOW	200
F/2 ?	RHS 120 BELOW	80
F/3	RHS 65 BELOW	20
F/4	RHS 220 BELOW	20
F/5 ?	RHS 65 BELOW	80
F/6 ?	RHS 300 BELOW	80
G/1	LHS 50 * BELOW	60
G/2	RHS 215 * BELOW	60
G/3	LHS 170 * BELOW	60
G/4	LHS 65 * BELOW	60
G/5	RHS 80 * BELOW	60
COL. 1 ?	CENTRE ? BELOW	50
COL. 2 ?	CENTRE ? BELOW	100
SIDE 1	0 BELOW	150
SIDE 2	0 BELOW	150

* INDICATES POSITION MARKED ON STRUCTURE WITH CHALK

Figure 9. Survey diagram showing location of cramps (GB Geotechnics).

becoming apparent. Having used the metal detectors to successfully locate, and mark some of the iron fixings the remainder of the data was gathered for analysis off-site. The locations of the fixings based on English Heritage predictions, would be provided by means of a table to give precise measurements from visible features and by indication on a drawing. Figure 9 gives distances in millimetres from the centre of the cramp to the nearest vertical joint between blocks within either the row of blocks 'above' or 'below' the horizontal joint. The approximate depth of the cramp beneath the surface is also given. The results of the resurvey suggested that the programme of installation proper should be reasonably straightforward. However, given the problems encountered in the trial it was considered that to ensure that no unidentified metal fixings in the structure were adversely affected by stray current corrosion the ICP system should be modified to localize the anodes in relation to known cramp locations as opposed to the original distributed anode design.

Modification of ICCP system

This design amendment was commissioned from Rowan Technologies Ltd, Manchester, who were already carrying out research for English Heritage into the corrosion of

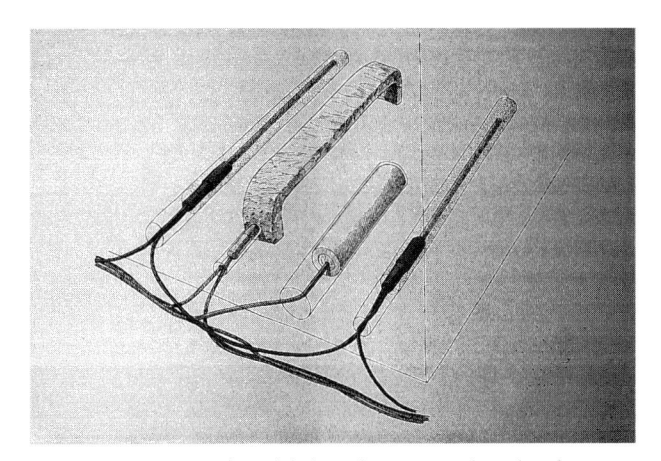

Figure 10. Cramp with modified anode system and single reference cell (English Heritage).

lead sheet roofing. The redesign used information gained from the various site surveys which suggested that cracking and damage to the stonework was localized mainly, not unexpectedly, around the cramps located close to the surface. It was the protection of these cramps which was to be addressed by the modifications in design.

The ICCP would now protect five cramps on the south elevation and nine on the north elevation, two of which were within the pilasters. Monitoring would be achieved by placing five silver/silver chloride/potassium chloride reference electrodes at selected locations. Two-piece anodes would be placed on either side of the cramps, approximately 50 mm away from them and parallel to their anticipated orientation (Figure 10). The primary anode would consist of a 3 mm diameter platinized titanium rod cut to 100 mm in length. This would be embedded centrally within a predrilled hole, 200 mm deep x 12 mm diameter, the hole backfilled with graphite paste. The paste would be injected into the hole and kept 20 mm short of the outer face of the stone allowing room for a lime-based mortar plug to fill the hole to the face. The same mortar would be also be used to embed the circuit cables. The new system layout is shown in Figure 11.

The DC negative and positive connections would be in a ring main configuration with cramp and anode connections being insulated with double layer mastic-filled heat-shrink sleeving. Individual reference electrode cables were taken back to the monitoring panel on the transformer rectifier, sited in the house.

Installation

With the development stages of the project completed full-scale implementation was planned to take place as part of an integrated package of repair aimed at a re-opening of the gateway in August 1995. The ICCP installation was the first phase of the package programme.

On site, while half the plotted cramps were located at the first drilling the others remained elusive. A cover meter revealed that the drill holes were in the vicinity of the elusive cramps but previous experience had shown that a large number of angled holes would have to be drilled from the original opening before access was achieved. In the trials this had then led to problems in locating the sleeve fixing system securely enough in the much larger diameter hole. A new method of final cramp location was required. The conviction that the failure to access around 50% of the targeted cramps was due to 'near misses' led to a design brief for a novel type of metal detector probe. The probe commissioned from GBG would be connected to a proprietary Imp pulse induction eddy current metal detector unit, made by Protovale (Oxford) Ltd, the type being used on site with the cover meter head (Figure 12). The probe, 5 mm in diameter and

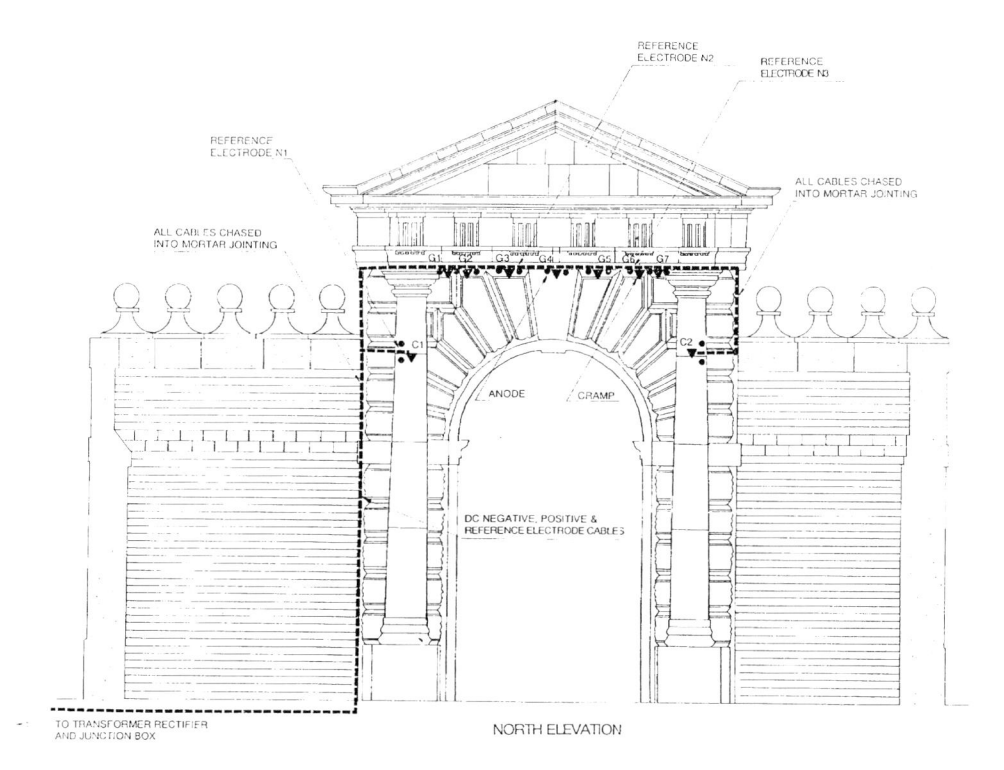

Figure 11. System layout (Rowan Technologies Ltd).

Figure 12. Directional probe developed for the Inigo Jones gateway project (English Heritage).

200 mm long, was fitted with a unidirectional detector head at its tip. Inserted into the first drill hole the probe was moved along and rotated at the same time. The metal detector gave an audible warning of metal within a 35–40 mm radius of the probe's detection head. With the cramp detected the probe was manipulated and the detector adjusted to give the precise position and depth of cover to the cramp. A new hole was then drilled to the cramp from the original point of entry. This technique proved to be effective in locating the remaining cramps. With all cramps located the ICCP system was installed. A number of the cramp connections were modified to a simpler self-tapping screw type where the deteriorated nature of the stone gave sufficient access to cramps close to the stone surface. Ring main wires from the cramps, anodes and return leads from the reference electrodes were fed through the masonry by way of existing mortar joints between stones and chases cut into the repair mortar render at a few points where extant joints were unavailable, the collected wires exiting the structure into an underground conduit. The conduit runs for approximately 60 metres to Chiswick House, where the wires resurface for connection to the power source, in this case a mains-fed TR unit. The Inigo Jones TR is discreetly situated behind an existing false door at the end of the building's link corridor.

Options to extend

The decision to restrict the ICCP system to the cramps identified was partly taken on the basis that the system as installed could be extended relatively easily. This may be needed to cover any other cramps which may reveal themselves by characteristic staining or incipient disruption of the stone surface at some point in the future.

To explore the practical difficulties involved in such a system extension the fixing point of the top rail of a set of railings was added to the circuit. The top rail, which is in the extreme east end of the south elevation flanking wall, was corroding and bursting the rendered brick in the immediate vicinity of the fixing. A spur to the circuit was added, including additional anodes and a reference electrode, and the fixing was successfully incorporated into the main ICCP system.

Figure 13. Taking a reference cell reading at TR (English Heritage).

Figure 14. The completed project with ICCP installed in the gateway (English Heritage).

Other repair works

With the ICCP system installed, visually checked and electrically secure the other works to the structure were initiated. The programme, designed by the then Architecture and Survey Branch of English Heritage and carried out by English Heritage's Historic Property Restoration organisation before its privatisation, was divided into two phases.

The first phase concerned repair to deteriorated render and repairs to stonework, including replacement of a small number of selected stones and rebuilding of the previously collapsed section beneath the pediment and the repointing of the holes and chases resulting from the ICCP installation. A most important piece of work included in this phase was the provision of a new lead sheet-covering for the roof of the pediment. The original lead covering and its temporary fibreglass replacement had apparently been stolen so the new covering would have to be securely fixed to the stone surfaces. It was also considered that both the lead and its fixings would need to be fully insulated from the stone or render of the structure in case the ICCP currents initiated corrosion on the underside of the lead. The lead was insulated by provision of a double layer of Mogat roofing felt covered with building paper. The fixings were made by providing resin patches in the render of the top of the pediment with fixings embedded in the resin and the lead bossed down over the cornice edge and clipped to secure. The second phase included cleaning, consolidation and sheltercoating of the stone surfaces. The final works were the hanging and redecoration of the wrought iron gates themselves. The cleaning of the stone and render surfaces was accomplished using a Microparticle system from Jackson and Cox Ltd. This is a dry air-abrasive method using aluminium oxide particles with a range of 17–120 microns. The blasting media is expelled from the nozzle at various angles or vortices onto the stone surface. The visual level of cleaning decided upon was a general reduction of the black sulphate layers. Following cleaning the stonework was sprayed with 40 coats of limewater in an attempt to consolidate the more friable areas. This treatment was followed by the application of a lime-based sheltercoat consisting of three parts lime to eight parts aggregate with addition of casein and formalin.

The wrought-iron gate

A double leaf wrought-iron gate had been fitted within the gateway arch. The gate had been removed some years before to be repaired and cleaned of corrosion, after which it had been painted with a red oxide primer and put into storage.

At the completion of masonry repairs the gate was to receive the remainder of the paint system and then re-hung. The metalwork was inspected prior to painting and in some areas light rusting was noticed erupting through the primer. Testing by wetting the metal's surface with deionized water and applying potassium hexacyanoferrate (III) test papers to the moist surface showed the presence of active corrosion cells. Consequently, a regime of cleaning was specified prior to re-priming and application of the remaining paint system. Flaking paint and rust were removed by careful rubbing down of the metal using *beartex* silicon carbide scouring wool and fine abrasive paper. The metal was repeatedly washed with clean fresh water and re-tested to remove soluble salts. When clean, a small test area was established to confirm the compatibility of the new paints to the existing primer. The new paint system consisted of a rust-inhibiting primer followed by a micaceous iron oxide build coat and a black gloss decorative coat. The gates were returned to the arch and fixed into the masonry with poured molten lead which, when cooled, was caulked home to produce a secure weather-tight fixing.

System commissioning and monitoring

ICCP can be classified as an active conservation treatment. Once the system is energized it is usually set to run continuously. However, study of concrete-protecting

ICCP systems has shown that after some time the area around the metal may be re-alkalized and the metal pushed to a level of passivity sufficient to warrant switching off the system. These possibilities will be evaluated as part of the future monitoring of the Inigo ICCP system. At the present time reference cell readings are taken on site at the TR to advise on the condition of the metal (Figure 13). A further refinement of this smart building technology currently being evaluated for the Inigo Jones gateway is the fitting of data loggers to continuously record the output voltage of the TR and cramp potentials in relation to climatic change and consequential masonry wetting and drying cycles.

Summary

The successful application of the cathodic protection at Chiswick has demanded a multi-disciplinary approach, combining practical expertise in metals and masonry conservation practice with materials science and modern technology transfer skills. The low levels of structural intervention achievable through the use of this keyhole surgery may provide an answer to many of the thorny problems associated with rusting ferrous metal in historic buildings, be they stone, brick or concrete. Perhaps the greatest challenge in historic masonry structures terms is providing ICCP for iron cramps embedded in lead. Theorectically possible, this area will require further research in order to ascertain both levels of effectiveness and potential preferential corrosion.

ENDNOTES

1. Charlton J, *Chiswick House*, (London: HMSO, 1958) 32.
2. Vitruvius, *Ten Books of Architecture*, (New York: Dover Press, 1960), 51.

BIBLIOGRAPHY

K. G. C. Berkeley and S. Pathmanaban, *Cathodic protection of reinforcement steel in concrete*, (London: Butterworths, 1990).

British Standards Institution, BS 5493: Appendix G: *Code of practice for protective coating of iron and steel structures against corrosion*, (London: British Standards Institution, 1977).

– , BS 7361: Part 1: *Cathodic Protection: Part 1. Code of practice for land and marine applications*, (London: British Standards Institution, 1991).

J. Charlton, *Chiswick House*, (London: HMSO, 1958). Vitruvius *The Ten Books of Architecture*, (New York: Dover Publications, 1960).

ADDRESSES

GB Geotechnics Ltd, Downing Park, Swaffham Bulbeck, Cambridge CB5 ONB; tel: 01223 812464, fax: 01223 812462

Taywood Engineering Ltd, R & D Division, Taywood House, 345 Ruislip Road, Southall, Middlesex UB1 2QX; tel: 0181 578 2366, fax 0181 575 4956

Rowan Technologies Ltd, Carrington Business Park, Carrington, Manchester M31 4ZU; tel: 0161 775 4648, fax: 0161 776 4518

Society for the Cathodic Protection of Reinforced Concrete, Association House, 235 Ash Road, Aldershot, Hampshire GU12 4DD; tel 01252 21302, fax 01252 333901

EQUIPMENT DETAILS

Anode wire supplied by Elgard Cathodic Protection, 36 Rue du Mont-Blanc, F-01220, Divonne-les-Bains, France; tel: (33) 50 20 30 55, fax: (33) 50 20 30 60

Graphite paste supplied by Birch & Krogboe, Hovedgaden 54, DK - 8220 Brabrand, Denmark; tel (45) 86 261311, fax (45) 86 26 16 12

Reference cells supplied by Silvion, 10 Boyne Rise, Kings Worthy, Winchester, Hampshire SO23 7RE; tel (44) 01962 882637, fax (44) 01962 882008

Cable and components supplied by RS Components Ltd, PO Box 99, Corby, Northamptonshire NN17 9RS; tel (44) 01536 201234, fax (44) 01536 405678

Transformer rectifier supplied by Dynamics (Bristol) Ltd, Walrow, Highbridge, Somerset TA9 4AN; tel (44) 01278 780222, fax (44) 01278 781824

Probe metal detector head supplied by GB Geotechnics, Downing Park, Swaffham Bulbeck, Cambridge CB5 ONB; tel (44) 01223 812462, fax (44) 01223 812462

Metal detector supplied by Protovale (Oxford) Ltd, Rectory Lane Trading Estate, Kingston Bagpuize, Abingdon, Oxfordshire OX13 5AS; tel (44) 01865 821277; fax (44) 01865 820573

Wrought-iron gate paint system: Primer: one coat Permoglaze Rust-Inhibiting Primer 67 microns wet film thickness Build coat: two coats Permoglaze Micaceous Iron Oxide 85 microns wet film thickness Finish: two coats Sikkens Rubbol AZ 65 microns wet Supplied by Akzo Coatings PLC, 135 Milton Park, Abingdon, Oxfordshire OX14 4SB; tel 01235 862226, fax 01235 862236

ACKNOWLEDGEMENTS

The authors would like to thank the staff of Taywood Engineering Ltd for their help in the development of the project including the initial ICCP design work, and Kevin Davies, of Rowan Technologies Ltd, for his help and support throughout the design amendment stage.

AUTHOR BIOGRAPHIES

Keith Blackney is a private conservator specializing in the conservation of architectural metalwork. He has worked widely in mechanical engineering, firstly in the development of construction plant and railway vehicles, and later in the prototyping of special purpose machine tools. He is a former member of the Architectural Conservation Team at English Heritage, working at the Architectural Metalwork Conservation Studio from 1988–97. His interests are in the adaptation and development of methods from other conservation disciplines and from industry for the non-destructive treatment of exterior exposed metals.

Bill Martin trained as a stone conservator and ran his own studio for nine years before moving to the Council for the Care of Churches where he was Conservation Officer. He joined the Architectural Conservation Team at English Heritage in 1989, and his responsibilities include project management of research into the decay and conservation of historic tile pavements and the evaluation of masonry consolidants. He also managed the English Heritage Architectural Metals Conservation Studio in Regent's Park, London until its closure in 1997 and coordinates technical advisory work for the Architectural Conservation Team.

Conservation of the lead sphinx at Chiswick House

KEITH BLACKNEY
37 Goss's Lane, Cheshunt, Herts EN7 5EG
BILL MARTIN
English Heritage, Architectural Conservation Team, 23 Savile Row, London W1X 1AB

Abstract

An eighteenth-century lead sphinx in the park surrounding Chiswick House, London, had been the subject of repeated vandalism. Requests for conservation advice to the Architectural Conservation Team (ACT) of English Heritage provided an opportunity to demonstrate the possibility of truly conservative repair techniques while enabling the continuing presentation of the sculpture.

Key words

Lead sculpture, x-ray diffraction, armature, monitoring

INTRODUCTION

In November 1991 the Architectural Metals Conservation Studio (AMC), part of English Heritage's Architectural Conservation Team, began a study of the conservative repair of lead statuary. One focus for this work was a lead sphinx, part of a collection of outdoor sculpture in the grounds of Chiswick House, an English Heritage property in West London.

The sculpture dates from around 1748, being one of three copies supplied to Richard Boyle, third Earl of Burlington, from the studio of John Cheere at Portugal Row, Hyde Park Corner, London. The other two sphinxes no longer stand at Chiswick, having been removed to Green Park in central London, where they now serve as gate guardians.

Cheere, who was first in partnership with his brother, Sir Henry Cheere, whose yard was near St Margaret's, Westminster, had taken over the Portugal Row yard from Anthony, a younger member of John Nost's family, in 1739. It seems most likely that Cheere also acquired moulds for garden sculptures at this time and was well placed to supply the extensive demand for cast sculpture for great country houses. Much of the style of this work was borrowed from known and popular classical prototypes but with certain stylistic re-emphases to lend a rococo flavour. Cheere's yard was described as being like a 'country fair or market, made up of spruce squires, haymakers with rakes in their hands, shepherds and shepherdesses... Dutch skippers and English sailors enough to supply a first-rate man-of-war' (Gunnis 1964, 99–100).

Figure 1. The sphinx in its original location in the grounds of Chiswick House, but raised for lifting, with a stone copy in the background (English Heritage). See also Colour Plate 13.

It was common practice for lead sculptures of this period to be painted and Cheere's sculptures in his studio were described as 'cast in lead as large as life and frequently painted to resemble nature' (Whinney 1964, 123). It is also known that lead sculptures of this date could often be painted to resemble stone, providing a cheaper alternative to the real thing when designs were to be reproduced. Ironically, the sphinx at Chiswick was latterly flanked by a pair of Portland stone copies (Figure 1) set beside the main path leading through the park away from the rear elevation of the villa.

The parkland surrounding Chiswick House is maintained by the London Borough of Hounslow and is open to the general public during daylight hours seven days a week. Relatively unsupervised access has exposed the garden sculpture to regular incidences of vandalism and it was one such incident involving the sphinx which initiated the original AMC condition survey.

SURVEY

The survey was commissioned by English Heritage's Central Architectural Practice on behalf of its London Region Properties in Care Group with the intention of assessing recent and longer-term damage and evaluating different proposals for conservation with particular reference to the public access situation in the park. It was quickly evident from the survey that the sphinx had deteriorated seriously.

Lead is highly malleable, ductile and soft, making it ideal for the production of finely detailed cast ornament such as garden statuary. However, it is also heavy, the heaviest of the common metals. The combination of these qualities means that lead sculptures cannot support themselves without an internal core and an either complete or partial armature, usually made in wrought iron. Lead also has no load-bearing capacity because it lacks tensile or compressive strength and is so soft that it can be easily scratched with a fingernail. As it has a relatively high coefficient of thermal expansion (three times that of steel), lead is subject to buckling and fatigue cracking caused by diurnal and seasonal temperature changes. Perhaps the best known of these thermally-induced alterations is creep, the continuous flow of the metal caused by a combination of stresses induced by thermal expansion and contraction and the weight of the lead leading to a permanent distortion (Gayle & Lock 1980). Naturally, garden sculptures which are fully exposed to the elements are susceptible to all these factors, especially given the limits to which the material is pushed to produce sculptural forms.

The sphinx was affected by numerous tears and cracks, particularly at obvious points of structural stress and had also been heavily marked by graffiti. The cracking of the lead surfaces, deriving from the deterioration processes described above, allowed ingress of moisture inside the sculpture where both the core material and the wrought-iron armature had been affected. In both cases these alterations are characterized by the creation of expansive products which exert pressure upon the internal surfaces of the lead by which they are encased causing further splitting and allowing further ingress of moisture and organic material.

The external surfaces of the sphinx exhibited a stable sulphated condition with an attractive glossy patina. However considerable damage had been caused by vandalism, by graffitti carved into the soft metal (some of which were 2mm deep), by the crushing of the front paws, the loss of small shreds of lead from these areas and denting of areas on the trunk and hind quarters. Tears around the saddle cloth were caused by the slump along the backbone of the sphinx caused by the weight of children sitting on the saddle.

Previous repairs had been made, in a grey cement-based mortar, the most noticeable being a roughly shaped prosthesis to the left haunch.

The sculpture was firmly attached to a Portland stone surbase which in turn was set upon a composite, Portland plinth in common with the other, stone, sphinxes. The plinths were sound and complete with only a few slightly

Figure 2. Underside of the sphinx showing the brick and mortar core with loose-fitted bronze stays (English Heritage). See also Colour Plate 14.

open joints. The surbase, however, exhibited a number of unsympathetic repairs to cracks emanating from underneath the sphinx. Their location suggested expansion of ferrous fixings.

It was evident from this initial survey that before the sculpture could be given any form of treatment more information on both its construction and structural state was needed. To this end it was decided to remove the sphinx to the AMC studios where controlled investigation and analysis could be carried out. The plinth/surbase joint was released with fine wooden wedges and the sphinx and surbase were raised onto timber bearers. Samson-type polyester continuous loop straps (lifting capacity 2000 kg) were drawn beneath the surbase to facilitate lifting with a truck-mounted Hiab crane for transportation back to the studio.

STUDIO INVESTIGATION

At the studio the sculpture/surbase unit was placed upon a 2.2m x 800mm x 800mm polystyrene block and turned through 90 degrees to lie on its side. Polystyrene chip-filled bags were used to cushion the sculpture. Examination of the underside of the surbase revealed the ends of four bronze location dowels fixed into the surbase with a dense cement-based mortar. The cracks in the surbase evident from the site survey corresponded with the position of these and it was clear that the bronze dowels represented a previous, undocumented intervention replacing the original ferrous fixings that had been the cause of the damage to the stone by rust expansion.

With the surbase appropriately supported, the mortar around the bronze dowels was carefully removed with a combination of Deroter micro drill and suitably sized tungsten-tipped quirks and the surbase was released from the sculpture to reveal the inner core material. The stone surbase was set aside for later conservation repair.

Having discovered (by way of the bronze dowels) that the sphinx had been the subject of at least one, probably twentieth-century, intervention, it was not a surprise to find that the core, once revealed, was not the expected foundry grog but rather a loose array of soft bricks and carbonated lime mortar. Three bronze stays were arranged across the belly area (Figure 2) but curiously were not fixed to the lead of the sculpture in any way, and were merely set loosely in the core rubble.

The left rear paw was fractured away from the main body, held in place by the cement prosthesis. The paw itself had been filled solid with poured lead and one of the four bronze dowels had been set into the lead. The other three dowels had been fixed by pouring molten lead into rough pockets made in the core.

As it was essential that the internal surfaces of the sculpture could be inspected, the loose core material needed to be removed. While this was taking place the sculpture was carefully monitored for structural deformation against a specially constructed panel erected behind the working area where the relationship between the lead surfaces and a graph pattern was closely observed from three set points. No deformation was observed and the structure proved to be self-supporting; therefore the entire core was removed save some plaster material in the head which looked as if it may have been part of the original core.

The plaster filling was solid but damp and was quite unlike any of the other material found in the main, later core. It was also characterized by a number of deep holes bored through the material, similar to those seen on stone surfaces where masonry bees had been active. A small sample of the plaster was taken for laboratory analysis. The left rear haunch prosthesis was loosened, lifted away and discarded.

With the core removed it was possible to examine the internal surfaces of the shell, the wall thickness of which varied from between 6mm in most areas down to 3mm in a few locations. The wall thickness was nearly twice the average where joints had been made and more than twice under the applied detail of the saddle cloth. The internal surfaces were undulating and pitted, probably resulting from contact with the original foundry grog. Little sign of the reddish lead oxide deposits sometimes associated with the reaction of lead to lime was discovered.

Examination of the outer surfaces under x20 magnification failed to detect any vestiges of paint or similar surface treatment and testing revealed that variations of surface colour were the result of two types of surface alteration product, a thin, whitish film of lead oxide and a thicker, darker layer of lead sulphate. Both these forms of natural patination are generally stable and resist further corrosion.

There was, however, a noticeable difference in shading of the patination in one or two areas close to larger tears in the lead surface. In an attempt to discover the cause of this phenomena four small samples of lead were taken for laboratory analysis.

While results of the analysis were awaited the sculpture was set back into its upright position and was checked by daily visual monitoring for any deformation of the shell. None was reported.

ANALYSIS

All sample analyses were carried out by English Heritage's Ancient Monuments Laboratory.

The plaster taken from the head of the sphinx was examined by X-ray powder diffraction and the main constituent was shown to be gypsum ($CaSO_4 \cdot 2H_2O$) with a small amount of quartz. As gypsum was regularly used as a core material in the lead casting process it seems certain that the plaster retained in the head dates from the time of manufacture.

The four lead samples, two of which came from the lighter shaded areas of patination and two from the surrounding more general areas, were mounted in resin, ground and polished to 1 micron, coated with carbon and analysed by X-ray analysis in a scanning electron microscope. Analysis showed that all four samples were of pure lead with levels of iron, tin and copper below the detection levels of the system (c 0.2%). There were no compositional differences between the samples. The cause of the shading difference between repaired areas and those not affected can only be put down to possible earlier attempts at repair having damaged and altered the base patina.

Figure 3. Touching-in of graffitti with earth pigment in Paraloid B72 (English Heritage).

PROPOSALS FOR CONSERVATION

Studio investigations had revealed that the sculpture had undergone at least one major repair intervention, perhaps at the time that its lead companions were removed to be replaced by the Portland stone copies. The condition of the Portland stone together with the carving style of the copies would seem to date them at around the turn of or early part of the twentieth century.

While the later core, due to its heterogeneous composition, had slumped badly and failed to support the outer shell satisfactorily the lead was of sufficient thickness, given the basic barrel shape of the main structure, to prevent complete collapse. However if the sculpture was to survive in its park setting then it would have to be given a new core and extensive repairs would need to be carried out to the shell, including casting of new parts to improve structural integrity and security and to prevent ingress of moisture to the new core material.

In similar cases elsewhere, lead sculptures have been essentially rebuilt with new castings, a new core, (possibly of expanded polyurethane foam), and a stainless steel armature. This work was often followed by repainting, either plain or naturalistic dependent upon the results of paint sampling, analysis and interpretation. But in this case, absolutely no evidence of paint could be found and the two sphinxes now at Green Park were similarly devoid of any traces of coating when examined under binocular magnification.

It was felt therefore that the degree of repair required to make the Chiswick sphinx sound, and the level of unavoidable interference with the extant patination, would be unjustifiable, especially as there was likely to be a continuing risk of vandal attack whatever repair methods were employed. The decision was thus taken to modify, at least temporarily, the sculpture's environment by moving it *inside* the villa and replacing it with a third stone copy to be carved by the English Heritage Stone Conservation studio, thus enabling as conservative a programme of repair as possible to the lead sphinx.

REPAIR PROGRAMME

The repair programme consisted of works to the crushed areas of the lead detail, the touching-out of graffiti and shading of patination, design and production of an adjustable inner supporting frame, repairs to the stone surbase and design and production of a device for monitoring any movement.

The crushed paws were eased back into shape with specially created wooden dollies applied either side of the affected area. While it was possible to recover the main shape there was a loss of some of the finer detail associated with the crushing. It was decided not to recreate this detail as it would have entailed cutting into the original material. The paws were therefore left in the correct shape but without the finished appearance of the back pair. The damaged patina around the crushed areas was touched in to match the surrounding areas with a range of earth pigments in a solution of Paraloid B72 and acetone applied with a fine sable artist's brush. The same method was employed for the blending in of the variations in patination in other areas where damage had taken place. Experiments were carried out filling some of the deeper carved graffiti with a microcrystalline wax and pigment mix but it was finally agreed with the curators that touching-in using Paraloid B72 and acetone again would be most appropriate (Figure 3).

Any remnants of previous mortar repair were carefully removed with scalpels and de-ionized water applied with a small stiff bristle brush. The entire shell had previously been washed inside and out with a low-volume pressure washer at 0.6 MPa (900 psi).

With the retouching completed the outer shell was given a light coating of microcrystalline wax dissolved in white spirit and which was lightly buffed when dry.

Figure 4. Stainless steel armature frame assembled and set to the minimum adjustment (English Heritage).

As the sculpture was to be displayed indoors and would no longer be exposed to the same levels of temperature variation, moisture attack and risk of physical damage due to visitor interference, and since it had proved to be of good structural stability, it was decided that there would be no need to fill the shell with a solid core. In addition, as the missing areas of leadwork, such as the section of left haunch previously made up with cement, were not to be replaced, a core fill would have been unsightly and would have prevented an almost unique opportunity to present a piece of cast lead sculpture with a permanent visual access to the inner shell space for monitoring its condition and for visitor interest. Instead a stainless steel, adjustable frame was attached to the surbase, providing the necessary structural support to the internal surfaces of the shell to allow the sculpture to be securely fixed down but allowing it to be removed again. The design requirements of such a frame were intriguing as the opening into the base of the shell was smaller than the inner flanks of the trunk against which any bracing fixing would have to be placed. The armature would have to be capable of size adjustments within the sphinx itself. A design study at the AMC studio eventually resulted in the production of a fully collapsible armature that could be tensioned and released using specially developed tools.

The armature (Figure 4), entirely constructed from stainless steel (316 grade), consists of four blocks of 38mm x 38mm x 60mm linked together with flat material using manual metal arc welding to make a flattened H shape. A hole is drilled through the length of each block to clear the shank of a No.12 gauge wood screw to enable fixing into a medium-density fibreboard (MDF) base. The upper 50% of each hole is opened up to 16.25mm diameter; the two nearest the head threaded 3/4 British Standard Whitworth (BSW). On each of the rear blocks a 7.9mm hole is drilled inside and parallel to the large holes. The smaller holes are threaded 3/8 BSW. Two curved ribs run laterally the length of the sphinx body. At the head the ribs have threaded 3/4 BSW spigots welded in position to make a 90 degree down turn: this allows the rib to be screwed into the block. At the tail end, L-shaped plates with slotted bases are fixed as the spigots. A threaded bracket is bolted to the top of each rib at approximately a quarter of the total distance from the spigot. An adjusting screw is threaded through each bracket and a lead billet is cast around the screw end on the outward facing side and an adjustment wheel on the inner side. As parts of the armature would be visible in the eventual presentation the surfaces were grit-blasted with chilled iron to give an unobtrusive, uniform satin grey appearance.

The initial survey had revealed that the stone surbase had a number of fractures emanating from the dowel location sockets and removal of the dense cement mortars used to fix the dowels had allowed the complete separation of corner sections of the base. It was noted that this was a recurring problem and that poor previous repairs had been made. Following the complete removal of these unsuitable repairs the separated sections were reintegrated by dowelling and cramping.

Holes 12mm in diameter and 40mm long were drilled in each of the broken faces with the alignment assured by marking around the first hole with small blobs of acrylic artist's colour and placing the broken joint faces together in their exact location so that when separated the acrylic paint had accurately marked the location for the next drilling. Keyed stainless steel (316) grade dowels 10mm diameter x 70mm long were introduced into the cleaned holes (Figure 5) in conjunction with polyester mastic (Technifil Vertical) and the moistened joint faces buttered with neat high calcium lime putty. Once offered up the sections were clamped on a flat surface to ensure level alignment until the dowel adhesive had cured.

To further strengthen the larger joints, stainless steel cramps were introduced to the underside of the surbase.

Figure 5. Stainless steel dowel inserted into a surbase section (English Heritage).

Cramp channels were cut across the line of the joint and sockets drilled out at each end for turn downs. The cramps were fixed in the same way as already described for the dowels.

As the sphinx would be reliant for structural stability on the integrity of the body shell it would be essential to closely and accurately monitor the form for any movement. Although the responsibility for the monitoring would initially fall upon the AMC team it seemed likely that at some point in the future this might be transferred to the house staff at Chiswick. Any regime therefore had to be both accurate and reasonably straightforward without the need for intricate, expensive equipment. However, the need for some equipment was clear and the overall design criterion was to provide an apparatus which was portable, accurate, stable and easy to set up and use from known datums. To achieve the stability and therefore the accuracy for repeated measurements the apparatus, a lighthouse type designed and produced at AMC (Figure 6), has a very low centre of gravity provided by a heavy base and column of mild steel with a solid lower section.

The base, which is designed to fit along the narrow top moulding of the new replacement plinth, is made in two parts held together with 3/8 BSW high tensile cap head screws. The two-part design ensures easier portability for the heavy base. The column has 1/2 BSW stud fitted at the lower end where it screws into the base section. To aid universal positioning from the narrow plinth moulding the column may be screwed into one of five threaded holes in the upper base block.

Figure 6. Lighthouse-type measuring device (English Heritage). See also Colour Plate 15.

Measurements are obtained from a 600mm long engineer's combination rule held in a slot machined into the side of a slider fitted over the column. The fine tolerance achieved by extremely accurate engineering between the slider bush and column prevents movement on the lateral axis which could affect comparative data collection. A clamp plate locks the rule in the desired position and measurements are taken from the rule based on positions identified by initial setup photographic record.

REINSTATEMENT

Upon return to Chiswick House the statue was installed in a newly designated sculpture gallery situated in the link

Table 1. Reference points for monitoring shell deformation.

REF	JUN 93	DEC 93	JUN 94	DEC 94	JUN 95	DEC 95	JUN 96
1L	234mm	234mm	233mm	234mm	233mm	233mm	233mm
2R	207mm	206mm	207mm	207mm	206mm	207mm	206mm
3	199mm	199mm	199mm	199mm	199mm	199mm	199mm
4L	160mm	160mm	160mm	160mm	160mm	160mm	160mm
5R	177mm	177mm	176mm	177mm	177mm	176mm	177mm
6	590mm	590mm	590mm	590mm	590mm	590mm	590mm
7	883mm	884mm	883mm	884mm	884mm	883mm	883mm
8	535mm	535mm	534mm	534mm	534mm	534mm	533mm
9	446mm	446mm	446mm	446mm	446mm	446mm	446mm
10	338mm	338mm	338mm	338mm	338mm	338mm	338mm
11	370mm	370mm	370mm	370mm	370mm	370mm	370mm

building to the main villa where a new facsimile of the original plinth had been constructed in timber and MDF and painted to resemble stone. The sphinx and the surbase were transported back to site as separate items, as the surbase would be fixed to the plinth by concealed fixings through the new armature.

The surbase was positioned on the plinth and the armature was positioned onto the surbase located into the sockets originally cut to accommodate the ferrous and later bronze dowel fixings. Pilot holes were drilled through the armature clearance holes into the MDF plinth and No. 12 gauge stainless steel wood screws fitted. The screws were tightened down until the screw heads reached the bottom of the counterbores and the surbase was clamped to the plinth.

The fixing frame adjustable movement was assembled onto the frame and collapsed by fully retracting the adjustment screws and moving the ribs toward an imaginary centre line through the stone. The sphinx was lowered over the armature and positioned correctly on the surbase. The ribs, which were accessible beneath the rear of the sculpture, were then opened out until the slots in the plates at the rear corresponded with the small holes in the blocks and could be bolted down.

Then a cranked 4mm diameter tool with a hooked end was fed through one of the holes in the rear of the shell and, guided by an endoscope, located onto the two screw adjustment wheels which were wound out until the lead billets on the screw ends made light but firm contact with the inside of the sphinx body, securely locating it to the surbase. The cranked tool was then withdrawn.

The detached left paw was secured to the base by setting the refitted bronze dowel into a dry sand and polyester resin mixture poured into the dowel location socket.

An interpretation board explaining the conservation programme is now situated on a wall close by the sphinx in the link building.

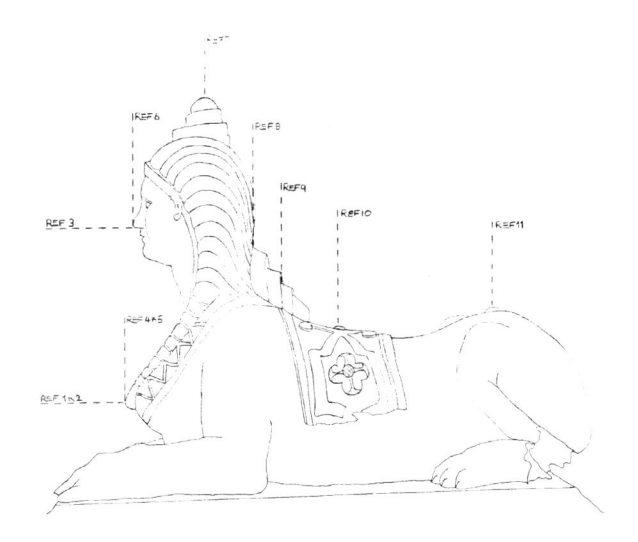

Figure 7. Examples of reference points for monitoring shell deformation (English Heritage).

MONITORING

As the method of presenting the sphinx, without an internal core, was a departure from the accepted procedure used in the conservation of lead statuary, it was considered essential to monitor the sculpture regularly to ensure that no irregular dimensional changes were taking place that could be related to deformation of the shell. The measuring

Figure 8. The lead sculpture, after conservation, on display in the Chiswick House link building (English Heritage).

instrument already described has been used successfully at Chiswick and the results are shown in Table 1. The measurements are taken from set reference points, examples of which are shown in Figure 7. The results so far show that the sculpture is completely stable and there has been no deformation whatsoever.

CONCLUSION

The Chiswick House sphinx provided English Heritage with an ideal subject on which to demonstrate conservative repair techniques on lead sculpture and at the same time allow the sculpture to remain on public display albeit in a modified environment (Figure 8). The solution is completely reversible and has had minimum impact on the object itself. Monitoring to date has shown that the sculpture is completely stable in its present environment.

The stone copy which replaced the lead sphinx was on the plinth in the park previously occupied by the lead sculpture for less than six weeks before its head was smashed off by vandals.

BIBLIOGRAPHY

M. Gayle and D. Look, *Historical survey of metals*, (Washington DC: The Preservation Press, 1980).

R. Gunnis, *Dictionary of British Sculpture 1660–1851* (London: Abbey Library, 1951 revised edition 1964), 99–100.

M. Whinney, M. *Sculpture in Britain 1530–1830* (London: Penguin, 1964), 123.

EQUIPMENT

Lighthouse-type measuring apparatus
Materials: mild steel with phosphur bronze bush to sliding collar
Two-part base comprising lower base, machine finish solid billet 75mm W x 88mm H x 202mm L, and upper base, machine finish solid billet 75mm W x 48mm H x 202 mm L
Column: turned and fine polish finish tube 38mm ID x 46.62 mm OD x1100mm H
Slider: turned and fine polish (bush), fine turned finish slider body
Bush: 46.72mm ID x 95 mm L eccentrically set into 73mm OD x 95 mm L body
Overall weight: 25 kg

ACKNOWLEDGEMENTS

The authors would like to thank Helen Hughes, Cath Mortimer and Sebastian Edwards, English Heritage staff, for their particular assistance during the course of this project. Also we should like to thank Rupert Harris, conservator, for his guidance in the early stages of the project.

AUTHOR BIOGRAPHIES

Keith Blackney is a private conservator specializing in the conservation of architectural metalwork. He has worked widely in mechanical engineering, firstly in the development of construction plant and railway vehicles, and later in the prototyping of special purpose machine tools. He is a former member of the Architectural Conservation team at English Heritage, working at the Architectural Metalwork Conservation Studio from 1988–97. His interests are in the adaptation and development of methods from other conservation disciplines and from industry for the non-destructive treatment of exterior exposed metals.

Bill Martin trained as a stone conservator and ran his own studio for nine years before moving to the Council for the Care of Churches where he was Conservation Officer. He joined the Architectural Conservation team at English Heritage in 1989, and his responsibilities include project management of research into the decay and conservation of historic tile pavements and the evaluation of masonry consolidants. He also managed the English Heritage Architectural Metals Conservation Studio in Regent's Park, London until its closure in 1997 and coordinates technical advisory work for the Architectural Conservation Team.

Development and long-term testing of methods to clean and coat architectural wrought ironwork located in a marine environment

The maintenance of railings at the Garrison Church, Portsmouth

Keith Blackney
37 Goss's Lane, Cheshunt, Herts EN7 5EG
Bill Martin
English Heritage, 23 Savile Row, London W1X 1AB

Abstract

The methods to clean and coat modern architectural steelwork are well documented, and specifications are usually drafted with reference to British Standards Institution (BSI) and International Standards Organisation (ISO) literature. However, little recorded work has been carried out to assess the appropriateness and effectiveness of the treatments when applied to historic architectural ironwork. This paper, based on a practical conservation project, looks at the desalination, surface preparation and comparative testing of modern coatings applied to historic architectural wrought iron situated in a marine environment.

Key words

Cleaning, wrought ironwork, marine environment, comparative testing, coatings and fillers, metallography

THE ROYAL GARRISON CHURCH

Introduction

The Royal Garrison Church is situated approximately 80 metres behind the sea wall at Old Portsmouth, close to the entrance of the modern harbour at Portsmouth, Hampshire, England.

A religious house has stood on the site of the Domus Dei (God's House), a hospice sheltering both the local poor and pilgrims from overseas bound for religious shrines in the south of England, since its establishment in about 1212 by Peter de Rupibus, then Bishop of Winchester. During the Dissolution the Domus Dei was surrendered to the Crown in 1540, functioning for some years after as a military store. Later in the sixteenth century Government House was built in front of the Domus Dei to accommodate Portsmouth's military Governor, and the former religious building returned to service as the Royal Garrison Chapel. On 21 May 1662 Government House was the venue for the marriage of Charles II to Catherine of Braganza, daughter of the King of Portugal. Government House was demolished in 1826, but the chapel continued to be used by the military for parade services. During the period 1866–9 a major restoration of the church was carried out under the direction of George Edmund Street, one of the leading exponents of Victorian church architecture. The nave, now a burnt-out shell, is a witness to more recent events when the church suffered severe incendiary bomb damage during the Second World War.

In 1933 the church was Scheduled as a building of historical importance and its upkeep became the responsibility of the Office of Works. The Royal Garrison Church is now in the care of English Heritage (EH).

Despite the high sea wall, which suggests some degree of shelter, the building is regularly exposed to extreme weather. It is common for strong, predominantly southwest winds to blow over the sea wall and vortex down around the church and environs: the local environment closely matches that described in BS 5493 as 'Exterior Exposed Polluted Coastal' (BSI 1977).

Ironwork

The site contains a large collection of ironwork. Gates and window ferramenta are fitted to the building, and groups of railings with gateways form the boundaries of both the churchyard and the two grassed fields situated directly behind and to the right of the building. The design of most of the ironwork suggests manufacture during the Victorian period. It is believed that the railings around the churchyard, the subject of this paper, date from the time of Street's restoration. Evidence to support this theory is given by a painting hanging inside the church, believed to have been completed around 1892, showing red-coated and pith-helmeted soldiers parading near the church, with railings of a similar design to those presently around the churchyard shown in the background.

Along the north, south and west boundaries the churchyard ironwork consists of railing panels made from square section palings piercing through flat-section top and bottom rails. Each panel is approximately 0.76m high by 3.4m wide and is joined to its neighbours with lapjoints formed at the rail ends. The panels sit over a low stone wall by means of fitted stanchions, on average two stanchions per panel, which locate into sockets cut into the wall's coping stones and are secured with caulked lead. Extra vertical support is provided by stay bars clamped to the stanchions with collars and held at their bottom in foundation stones set into the ground approximately 300mm behind the wall. The west boundary wall is bisected at mid-point by a double leaf gateway, with an approximate opening of 1.8m, which serves as the main entrance to the churchyard. Along the east boundary the railing design is modified to accommodate a rise in

Figure 1. Gateway and railings at Garrison Church. See also Colour Plate 16.

ground level: the bottom rail is omitted and the palings are longer and individually fitted at the bottom into a stone curb sunk in the ground.

The collection's overall design is functional rather than decorative and embellishment is restricted to simple scrollwork on the gateway and trident finials fitted to the railing panel stanchions (Figure 1).

CONDITION IN 1993

Survey

A condition survey of the churchyard ironwork was carried out during February 1993 at the request of English Heritage. The ironwork's maintenance had followed the usual pattern of successive painting regimes, one over the other, gradually building up to a thick coating which masked finer points of detail and methods of construction. The last maintenance appeared to have been some time ago as the black gloss decorative finish had dulled considerably and the coating was now cracked and flaking. Consequently little protection was given to the underlying metal which was, in many places, in an advanced stage of corrosion. Where the paint had fractured it could be seen that, until recently, the coatings were coloured rather than the ubiquitous black commonly associated with exterior ironwork.

While still functioning the gates suffered a number of mechanical defects which contributed to their overall decline. Wear of journals, and possibly some ground settlement, had left the gates out of alignment and prone to jamming when shut. The key-operated lock had seized in the unlocked position and the gates were now secured by a padlock and chain. A curious aspect of the gates were the very obvious welds at every joint between individual components. The thickness of the paint made their function difficult to determine but the welds appeared to be in addition to, rather than substituting for, the traditional blacksmith's method of pierced and riveted joints normal on ironwork made before the early part of this century.

On both gates and railings, at least 50% of the metal's surface had been oxidized to firmly-adhered scales of hard corrosion. It was particularly noticeable that most corrosion had occurred on the top surfaces of rails and on the south and west faces of vertical bars. The corrosion was greater in volume than the iron and had a distinctive laminar appearance, suggesting that the parent metal was wrought iron. Although generally well adhered the corrosion had not sealed the metal from further attack. Moisture appeared to permeate between the interface of metal and corrosion with the resultant run-off transformed to a light brown rust-coloured solution which flowed over painted surfaces and masonry, leaving them heavily stained. Most of the railing panel stanchions had corroded where they entered the coping and the expansive corrosion product had, at least in part, been responsible for bursting the surrounding stone (Figures 2 and 3). The weakening of the stone had resulted in considerable destabilization of the overall structure and a number of panels now relied on their stay bars and neighbouring panels for support. This was particularly noticeable for the group of railings along the west boundary.

Figure 2. Disruption of coping stone due to expansion of rusting stanchion. See also Colour Plate 17.

Figure 3. Hard scales of corrosion. See also Colour Plate 18.

It was clear that the ironwork was in an advanced state of deterioration and the prime mechanism of decay, the corrosion, was far greater than that expected from exposure to oxygen and moisture alone. The aggressive corrosion was thought to be a result of the metal's chemical contamination with salts from the local marine environment. Given the railings' instability along the west boundary, and the need to carry out repairs to the stone coping, it was decided to remove and repair this group of ironwork at the former English Heritage Architectural Metals Conservation Studio (AMCS). Upon reinstatement the group would be regularly monitored and the collected data used to inform treatment of the remaining ironwork.

PROJECT BRIEF AND PREPARATION

It was decided that repairs to the iron would be limited to those required to return structural stability, and that no attempt would be made to disguise the effects of decay on purely aesthetic grounds. Research would be carried out into available methods of identification of harmful contaminates, and testing the practical application of a range of cleaning systems, their cleaning effectiveness and effect on sound material. A range of commercially-available surface coatings for long-term comparative testing in the punishing environment at the Royal Garrison Church would also be selected. The procedures and treatments for testing needed to be sufficiently technically developed for use within the time-scale of the project, and financially viable for application in similar future cases. The AMCS team were keenly aware that, with a few exceptions, for example the Practical Building Conservation Series (see Ashurst 1988), there is a lack of literature on the treatment of historic architectural ironwork. Specifiers often had to turn to mainstream construction engineering standards for guidance in drafting conservation specifications. The team decided to examine the most regularly quoted standards literature to assess their appropriateness in terms of both the Royal Garrison Church project and future conservation applications.

The church railings were to be reinstated by the end of May 1994, in time for the 50th anniversary of D-Day. With existing commitments this deadline meant that the project would have to be completed within two months of its inception. It was against these criteria that the ironwork was removed in December 1993 to the AMCS for studio work to start in March 1994.

Dismantling

Before dismantling all elements were numbered and catalogued. The gates were supported while the gate-hanging stile to gatepost-retaining clasps were released. The pivots at the bottom of the gate-hanging stiles were lifted from the iron pods set in the stone threshold and the gates were removed for transportation. The lead caulking was loosened and the pods extracted from their stone sockets. The railings were supported and rivets removed from inter-panel lap joints. Stanchion stay-bar stone foundations were excavated and the lead caulking removed to free the stays. The lead caulking in the coping around the stanchion bases was already loose enough to allow the panels to be removed. Individual panels were lifted away from the stonework for transportation. A limited amount of further dismantling was carried out at the studio, unfastening rivetted pins and removing collars to allow the separation of stay bars from stanchions.

MATERIALS TESTING

The lack of firm archival material to date the railings, the obvious changes in decoration and peculiarities in the gate's construction prompted the commissioning of a series of laboratory investigations to be carried out by the English Heritage Picture Conservation Studio (PCS) and the Ancient Monuments Laboratory (AML).

Paint analysis

Selective sampling of painted surfaces was carried out by the Architectural Paint Research Section (APRS) at the PCS. The samples were analysed to establish a chronology of decorative finish and to advise on the re-decoration of the ironwork.

The samples were cast into resin blocks and sanded with progressively finer abrasives to produce the flat polished surface required for accurate microscopic

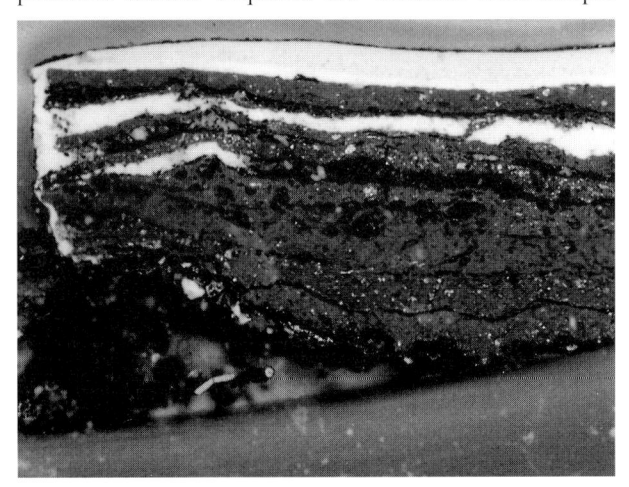

Figure 4. Paint sample under 50x magnification. See also Colour Plate 19.

Figure 5. Gate heel showing fabricated strengthening blocks.

examination. 50 times magnification showed the samples to consist of many more layers of paint than had been observed during the site survey (Figure 4). The layers appeared to have resulted from three distinct decorative periods. The two later phases have smooth, flowing, regular thickness strata consistent with twentieth-century paints, the lower layers were made up from a spectrum of greens, while the most recent painting had produced three complete systems consisting of primer, undercoat and a black decorative finish.

Below the modern paints were 12 layers exhibiting a range of brick red colours and characterized by rough uneven strata. The APRS had earlier discovered similar reds on exterior iron railings at two separate sites in London: Caledonian Market, *c* 1850 and 43 Whitehorse Road, *c* 1875. Initially there was some doubt as to whether the lowest red paint was part of the original coating system as it overlaid a substrate of thick corrosion. However another sample revealed the same shade attached to a millscale substrate. Millscale is a thin oxide skin formed on the metal's surface during air cooling from the high temperatures employed in the material's production or subsequent forging. Although offering a barrier against the environment unpainted iron left to weather will rapidly detach millscale. It was concluded that this shade of red had been applied soon after forging, and was part of an original paint scheme.

The new decorative colour was determined by comparing the samples against various manufacturers' colour swatches until a match was made with CO.1O.2O, specified in the Sikkens Colour Collection 2021 range of paints.

Metallographic investigation

Despite the paint analysis some uncertainty remained as to the age of the ironwork. The gates in particular exhibit some characteristics inconsistent with nineteenth-century manufacture. Gates made by the blacksmith using traditional techniques normally feature mortice and tenon joints to join frame components. To form a tenon with a high ratio of height to thickness, and a structurally sound joint, the rails are often increased in vertical section, 'upset', where they join on to the hanging stile. In the case of the Royal Garrison Church railings, inspection after cleaning showed that the welding, often crudely executed, was generally the only method of joining the frames' individual components. The rails had not been upset but instead fabricated to appear upset by adding crudely fashioned strengthening blocks welded into position (Figure 5). Furthermore, on both gates the lower journals were made by holes drilled longitudinally into the bottom of the hanging stiles into which overlong plugs had been fitted and left protruding to act as pivots. This design is sometimes present on traditionally-made ironwork, but in this case it is at odds with mortice and tenon construction. The bottom rail is mounted so low down on the hanging stile that the pivot holes would severely limit the amount of tenon passing through the stile. However some aspects of traditional blacksmith construction are present. The vertical bars are made from continuous lengths of material which pass through holes made in the locking and top rails. At the bottom they pass

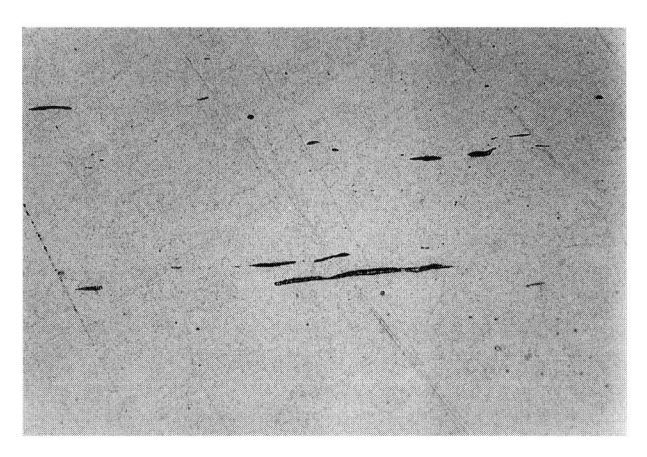

Figure 6. Micrograph: unetched x400. Slag inclusion elongated along length of bar.

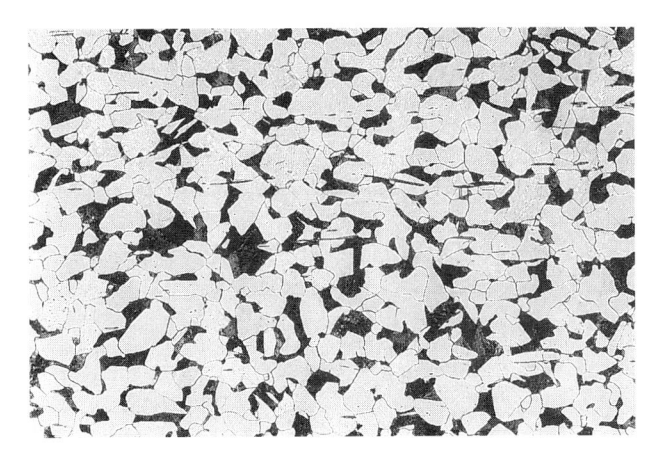

Figure 7. Micrograph: etched in 2% nital x200, ferrite and pearlite.

through the bottom rail, but are secured by welding rather than the expected rivetting over. A further sign of traditional construction was the discovery, buried within the left gate hanging stile, of the stub end of a tenon which did not appear to be connected to the adjacent bottom rail.

The Ancient Monuments Laboratory was asked to examine the metalwork for evidence that would date the materials and construction. Small metal samples were removed: sample 1, a horizontal cross-section, which was sub-divided vertically, from a paling on railing panel 4, and sample 2, cut at an angle incorporating sections from the hanging stile, bottom rail strengthener and weld metal of the right-hand gate. The samples were mounted in conductive phenolic resin and prepared by sanding with progressively finer abrasives to produce a mirror-like finish. The finished samples were examined under 400 times magnification and recorded (Figure 6).

The samples were further prepared by etching with a 2% nitric acid in alcohol solution and re-examined under 200 times magnification (Figure 7). A Shimadzu microhardness tester was used to determine the hardness of different phases within the metallographic structure.

The metallographic tests showed both the railing and gate samples to be made from wrought iron, albeit with an unusually high carbon content, approximately 0.2% for the gate. The level of carbon is almost sufficient to define the metal as a carbon steel, and is an unusual choice of material in this application. The very even distribution of carbon within the iron suggests it to have been made using the puddling process, first developed by Henry Cort around 1784 and thereafter in continuous use in the United Kingdom until the closure of Thomas Walmsley's plant in 1973. At the time of writing limited wrought iron production had, intermittently, resumed at Blists Hill, part of the Ironbridge Gorge Industrial Museum complex.

The composition of the weld metal and the comparatively narrow heat-affected zone through the wrought iron suggests that the welds were made by the electric manual metallic arc (MMA) 'arc' or 'stick' welding process. The general commercial use of MMA began around the early 1890s, some 20 years after Street's restoration of the church railings.

Conclusions

Taking into account the painting hung within the church and the findings of paint and metals analysis the weight of evidence tips toward at least some of the ironwork dating from the middle or late Victorian periods, possibly from Street's restoration.

The millscale find strongly suggests that the paint is an original coating and the paint analysis, cross-referenced against other APRS casework, provides the strongest link with the mid Victorian period. The metallographic tests were useful in confirming the railing and gate samples as wrought iron and the discovery of a high carbon content was most intriguing.

The suspicion that the ironwork may date later than the 1860s arises solely from the use of MMA welding and the fabrication of the rail ends. However, closer examination shows the gates to have at least some aspects of traditional construction, for example the rails pierced through to accept vertical bars. The redundant tenon stub indicates that traditional construction of the frames was also employed at some stage. Consequently it is suspected that the fabrications and welding result from a later, major, restoration of the gates rather than from their original manufacture.

The mystery of the high carbon content in both the apparently original and repair metal may be explained by the Royal Garrison Church's strong links with the local military. Street's restoration was funded in part by both the Army and Royal Navy. The local Royal Navy dockyards are likely to have held stocks of carbon steels and had available MMA welding technology from an early date. It can be hypothesised that the ironwork, and/or its repair, has a military provenance.

IDENTIFICATION AND TREATMENT OF SOLUBLE SALT-INDUCED CORROSION

Testing for chemical contamination.

Despite the corrosion being generally well-adhered, by careful chipping with hammer and chisel it was possible to remove it from a number of selected areas on two of the railing panels and the right-hand gate. In all cases the corrosion followed the same distinct pattern. Hidden beneath a comparatively even outer surface the corrosion had penetrated to a variable depth characterized by a series of shallow-sided craters. Each crater had a pronounced eye of increased attack at the base. The excavated test areas were brushed free of remaining loose particles of corrosion using a stiff nylon brush and tests were carried out using two complementary techniques.

Potassium Hexacyanoferrate (III) ($K_3Fe(CN)_6$)
Potassium Hexacyanoferrate (III) ($K_3Fe(CN)_6$) papers were used to test for soluble salts of iron (BSI 1977, Appendix G; Haigh 1993). The conservators were unable to find a commercial supply of ready-to-use papers, so those used in the test were made at the studio by immersing 125mm diameter filter papers into a 5%

$K_3Fe(CN)_6$ deionized water solution (Ashurst 1988). When dry, the test papers were pressed against the test area surface, which had been wetted with a fine spray of deionized water. The contact areas of the papers immediately turned a dark blue indicating the presence of soluble salts of iron: the shallow sides of the crater appeared as a matrix of blue cells increasing in concentration to become a complete covering at the eye of the crater.

Silver nitrate
Tests using silver nitrate were carried out to define whether chlorides were present, and consequently responsible for the precipitation of the salts of iron found in method 1. Deionized water was poured into the craters in an area of erosion and the test area sealed over with cling film. After standing for approximately 8 hours the water was collected into a clear glass container and silver nitrate solution added. The water turned a cloudy white indicating the presence of chlorides.

The $K_3Fe(CN)_6$ test paper method provided an easy and reliable method of detecting contamination and was adopted as the project's principal method of testing. Other methods to test for chemical contamination were assessed including those in BS 7079 (Haigh 1993), but they were considered too complex, time-consuming and potentially costly when assessed against the project criteria.

COMPARATIVE TESTING OF CLEANING SYSTEMS

The detection of soluble salts and active corrosion beneath the scaly corrosion confirmed that removal would need to be attempted. The continued presence of such contaminants would severely restrict the options for, and success of, cleaning and coating methods. Earlier projects carried out by the AMCS team to treat cast iron from a similar coastal site, Osborne House on the Isle of Wight, had shown that dry air abrasive cleaning alone would not remove soluble salts from the metal's surface. The iron, despite being cleaned of all visible corrosion, would rapidly re-corrode due to the salts' hygroscopic nature and propensity to create a low pH condition on the metal's surface. Consequently for the Royal Garrison Church project it was decided to test a number of different cleaning systems, including the soaking technique used for the Osborne project, to establish the most effective, practical method of surface cleaning.

The cleaning trials were carried out on two panels and the right-hand gate. Five cleaning methods were chosen: three with later AMCS modifications, described in BS 7079 Part A1, and two others also used in industry to clean steelwork.

BS 7079 Part A1 is part of an international standard, ISO 8501-1: 1988 (E). The standard is based around 28 photographic examples depicting four metal surfaces at varying levels of corrosion before and after treatment using three alternate methods of cleaning.

Method 1: BS 7079: Part A1. Hand and power tool cleaning; designated St
A range of hand and mechanical cleaning methods were applied. To avoid damaging the metal surface, the grinding and needle gun techniques in the BS document were excluded. The work was labour-intensive and time-consuming. Difficulty was experienced in achieving St 2, the lowest level of cleaning recommended for painting, which was in itself insufficient to expose the iron surface for thorough salts testing.

Method 2: BS 7079: Part A1. Flame cleaning; designated Fl
The corrosion was heated using a oxy-acetylene torch variously fitted with a range of welding and heating nozzles and brushed both manually and mechanically with steel wire brushes. This proved to be speedier than the St method and satisfactorily exposed the iron for salts testing. However, a thorough clean was still difficult to achieve without concentrated and time-consuming effort. Particular attention had to be given to the health and safety issues raised by the fumes of burning paint and the rapid, directionally unpredictable, detachment of hot corrosion product.

Method 3: BS 7079 Part A1. Blast-cleaning; designated Sa
A panel with two stanchions was selected for testing. Prior to blast-cleaning the metal was de-scaled of heavy corrosion by flame and hand cleaning, and hot water pressure-washed to remove dirt. The metalwork was allowed to dry naturally and was placed in a shipping container converted into a sealed air-abrasive blast room by Powerblast Ltd. Cleaning was carried out using recyclable G17 grade chilled iron grit at a range of test pressures no greater than 5.5 bar to achieve a surface profile of 50-70 microns and a standard of cleanliness to match Sa 2.5. The metal's surface appeared a bright silver/grey colour clear of any residues that might conceal areas of salt contamination. One of the stanchions was left in the blast room overnight, and when examined the next morning the areas previously identified by the test papers as contaminated had produced a dark brown to black rusted surface (Figure 8).

METHOD 3A: BLAST-CLEANING FOLLOWED BY HOT WATER PRESSURE WASH
Immediately after cleaning the panel was repeatedly hot water pressure-washed, at 100 bar nozzle pressure, with mains water heated to 30° and tested until indication of salts had largely been eliminated.

METHOD 3B: BLAST-CLEANING FOLLOWED BY IMMERSION IN UNHEATED DEIONIZED WATER
The stanchions were immersed and soaked in a tank of deionized water. Readings were taken using a WPA CMD 630 conductivity meter fitted with a 2022/670 model electrolytic conductivity cell and the water changed until the rise in conductivity stabilized at a consistent, low level measured in parts per million per hour, between water changes. The bars were then removed from the tank and hot pressure-washed to remove the orange bloom of loosely adhered hydrated oxide that had formed on the metal's surface. Finally $K_3Fe(CN)_6$ tests were carried out to confirm removal of salts.

Figure 8. Potassium Hexacyanoferrate paper test indicating presence of soluble corrosion salts immediately after blast cleaning.

Method 4: Blast-cleaning with entrained water
BS 7079 does not include visual standards for surfaces after wet air abrasive blast-cleaning, however the document notes the necessity to remove salts, and recommends that wet blasting be used in these cases. The document acknowledges that wet blasting may produce inconsistencies of colour between individual cleans but suggests the Sa images may still be used to give an indication of preparation grade. Corrosion inhibitors are sometimes incorporated into the water stream to avoid flash rusting, 'gingering': inhibitors were rejected for the Royal Garrison Church project to avoid any possibility of residual recontamination of the metal. An alternative method of drying the metal's surface using the air stream from the blast gun was tested.

The wrought iron was hand-cleaned similar to preparation for the dry abrasive process. Blast-cleaning was carried out in a specially constructed plastic-sheeted enclosure; run-off water was filtered to remove spent abrasive and removed corrosion, paint etc. Blasting with a range of test pressures was carried out using a Hodge Clemco Ltd contractors' type portable blast unit and J Blast expendable abrasive.

Testing after blasting showed some salts still present and hot water washing was then carried out until the level of salt readings reached those achieved by the blast-cleaning followed by the hot water pressure wash. The prevention of gingering by using the air blast was only partially successful as the first sections of metal cleaned tended to begin flash rusting before the clean was completed and the drying could begin. The main advantage of the wet blast compared to dry blasting is usually cited as the easier containment of the expended grit and debris removed from the iron. However, this is only accurate in terms of site practice. A properly-equipped dry abrasive facility, such as the Powerblast room, provides far higher levels of safe containment.

Method 5: High pressure water jet cleaning
A Harben mobile pump was hired and set up for operation within the sheeted area in the AMCS compound. The pump can deliver fine jets of water at nozzle pressures approaching 1000 bar, and the attendance of a jetting expert was required to ensure safe operation of the equipment. Using a rotating head lance, the pump pressure was gradually increased: removal of the corrosion and loose paint was proportionate to the increase in pressures. By approximately 800 bar the corrosion was being removed to leave a clean, yet unworked, metal surface. Interestingly, even at this high pressure, where the paint was sound it tended to remain adhered to the metal unless the lance was concentrated over the area for some time. Salts testing showed a marked decline in salt concentrations after a comparatively short clean time. Similar to the wet abrasive method high pressure water presented problems in containing the debris removed from the iron. There was one, unexpected, aspect of water jetting that reduced the usability of the technique. As pressure increased the lance became more and more difficult to control; a fixed stance had to be taken by the operator and by 800 bar even the more muscular of the AMCS team began to have difficulty in moving the lance without being lifted bodily from the ground!

Final selection of cleaning method

After assessing the results of the cleaning tests it was concluded that, in this case, an initial mechanical and flame clean to remove bulk corrosion followed by dry air abrasive cleaning and either hot water washing or soaking in deionized water was the most appropriate treatment in terms of controllability, cleaning efficiency and cost-effectiveness. It was realised that the flash rusting resulting from washing might limit the choice of coatings, but this would be dealt with by either applying coatings tolerant to lightly rusted surfaces, albeit ones cleaned of chemicals such as chlorides, or by flash-cleaning the metal with dry air abrasive.

It was recognised that air abrasive cleaning would rework the metal's surface and consequently might not be acceptable for objects where surfaces retained original detail or other information of historic significance. However, in the case of the Royal Garrison Church project it was considered that the loss of original surface, or what little remained, would be offset by other long-term benefits.

SELECTION OF COATINGS AND FILLER

Coatings

A number of factors had to be taken into account in the selection of suitable coatings to be included in the trials. Firstly, the long-term performance of the coating was directly related to the metal's surface condition immediately before application. Secondly, the selected coatings should include systems already incorporated in specifications to treat historic wrought iron similar to that at the Royal Garrison Church. Finally, the coating should be both durable and simple to maintain within a limited maintenance budget.

Two coating systems were developed with reference to BS 5493. This document contains information relating the required preparation of metal surfaces, composition and thickness of coatings, and projected performance, termed 'years to first maintenance', of a variety of coating systems in a range of environments. Table 3. Part 3. *Exterior Exposed Polluted Coastal Atmosphere* was selected as being most applicable to the conditions at the Royal Garrison Church.

A third system, not included in the BS document, was selected, for two specific reasons. Firstly, its suitability for application over rusted surfaces, albeit in this case 'clean' rust consisting only of ferrous hydroxides. And, secondly, to acknowledge the growing demand both by legislators and consumers for 'environmentally friendly' paints that do not depend on the evaporation of non-water solvents as part of the drying process.

System 1
- Substrate: flame sprayed zinc $\cong 100\mu$ thickness
- Primer: two coats Sikkens METACOAT P132 $\cong 70\mu$ wet film thickness
- Undercoat: one coat Sikkens ONAL $\cong 60\mu$ wet film thickness
- Finish: two coats Sikkens RUBBOL AZ $\cong 65\mu$ wet film thickness

This system closely resembles SC10Z and is expected to give 10 to 20 years protection before first maintenance. The system had been used successfully on earlier AMCS projects.

System 2
- Primer: one coat Permoglaze RUST INHIBITING PRIMER $\cong 67\mu$ wet film thickness
- Undercoat: two coats Permoglaze MICACEOUS IRON OXIDE $\cong 85\mu$ wet film thickness
- Finish: two coats Sikkens RUBBOL AZ $\cong 65\mu$ wet film thickness

Based on SF8 the system has an expected life to first maintenance of five to ten years. The system was specifically selected as one often drafted for the coating of historic architectural ironwork.

System 3
- Primers: one coat Acrylon LB 1O $\cong 38\mu$ wet film thickness, followed by one coat Acrylon MB 1O $\cong 54\mu$ wet film thickness
- Undercoat: one coat Acrylon MIDCOAT $\cong 200\mu$ wet film thickness
- Finish: Two coats Sikkens RUBBOL AZ $\cong 65\mu$ wet film thickness

Due to the cost and time required to mix a small quantity of a non-standard shade and after consultation with the primer and undercoats manufacturer it was decided to decorate this group with the same alkyd finish used for Systems 1 and 2.

Filler

It has been common practice to pack gaps between poorly fitted sections of ironwork with a filler, often a vegetable oil-wetted putty containing red and, historically, white lead. This type of filler had previously been used on the lap joints between the church railing panels, and, typically, was now found to have dried and cracked. Assessing the protective qualities of red lead putty once the oil had dried was beyond the scope of the current project. However, it was concluded from observations of the metal's condition, both on the Royal Garrison Church iron and concurrent AMCS projects, that protection was reduced once the putty had cracked and the barrier between iron and environment broken.

An alternative filler was sought which could be easily moulded and packed into gaps and water traps. The material would need to maintain elasticity and the ability to absorb any movement of metal. It should withstand exposure to the extremes of winter; and, in the heat of summer, remain sufficiently solid to avoid deformation and sagging. Densoseal 16A, a non-setting polybutene-based mastic containing organic fibres, used in industry to seal cable ducting against moisture ingress, was selected for the long-term test programme.

TREATMENT PROGRAMME

All components

Prior to repairs the metalwork was cleaned using method 3a; smaller components were treated using method 3b.

The ironwork was divided into three groups: group 1, the railing panels to the left of the gates; group 2, the gates and group 3, the railing panels to the right of the gates. Thereafter the treatments became group-specific.

Application of coatings was made with reference to BS 5493; BS EN 22063 (BSI 1977 and 1994), and BS 3900 (BSI 1994). Due regard was given to paint manufacturers' instructions concerning the parameters of temperature and humidity, and particular attention was given to achieving the correct thickness of coating by using Elcometer Ltd wet and dry film thickness gauges.

Individual groups

Group 1
A small number of consolidation repairs were carried out to panels 2 and 3 where the inter-panel lap joints had become severely wasted through corrosion. Consolidation was achieved by building up a series of overlapping welds with metal-active gas-shielded (MAG), 'MIG' or 'CO_2' welding. Both of the left-hand gate standard to railing panel connection rivets located at the right-hand ends of the rails of panel 4 were found to be extremely corroded, with the top rail rivet being 75% fractured through at its base. Both rivets were removed and replacements bronze-welded into position.

The panels were again washed and tested for salts. When dry, the metal was flash dry-air abrasive-blasted to remove the hydrated oxides formed as a result of washing. All surfaces were blown down to remove dust with clean dry compressed air, and tests were carried out to confirm the absence of grease. A 100μ coating of metallic zinc was applied using a Metallisation Mk 2 oxy-acetylene fuelled flame spray pistol. Application of paint was began within two hours of flame spraying.

Group 2
Screw fixings were released and the gate handles and lock removed. A small number of indent repairs were carried out to the handle sconces. Three handle rings were made in place of the missing originals. All new components and repairs were date-stamped. The lock was cleaned, lubricated and a new key made. The gates were again hot water pressure-washed and tested for contamination. When completely dry the metal was wire-brushed to remove loose powdery hydrated oxides formed as a result of washing and paint system 2 applied.

Group 3
A repair was carried out to a forge weld on panel 7 where the panel makes a 90 degree return to follow the masonry wall. The weld was of limited success and only a small area of combined material formed, this had torn the fibrous metal for a short distance before breaking. The sections were realigned along the fractured fire weld and MAG-welded together. Both the right-hand gate standard to railing panel connection rivets, located at the left-hand end of the rails of panel 5, were found to be corroded almost completely through. Both rivets were removed and replacements brazed into position. The panels were again washed and tested for contaminants and painting of system 3 began.

MASONRY REPAIR

During the absence of the metalwork, a programme of masonry repairs had been carried out. This centred around the indent repair of the location sockets cut into the stone copings, and the reconstruction, using stainless steel dowels and polyester resin grout, of the right-hand gate standard foundation stone. The repairs were carried out by craftsmen of the former English Heritage Historic Properties Restoration (HPR) Ltd.

REINSTATEMENT

All components were braced with wooden battens, wrapped in polyethylene bubble wrap and transported back to site.

Reinstatement of the metalwork was the reverse of the dismantling process. Molten lead was poured into the stone sockets around stanchion bases, stays and gate standards. When cold and solidified, the lead was caulked using hand tools to ensure a secure fixing. Inter-panel lap joints were re-rivetted using malleable iron rivets set cold to avoid burning the paint, and gaps between laps were packed with mastic. Gate standard to panel butt joints were re-riveted. The diameter and material of the rivet necessitated heating with an oxy-acetylene torch. Before heating, surrounding areas were masked with solder mats to minimise burning of paint. The pods were refitted and secured with caulked lead into the stone threshold, all journals were lubricated with Castrol CL lime-rich water-resistant grease and the gates re-hung. Finally, small areas of paint burnt or chipped during reinstatement were made good by sanding back to a sound feathered edge and reapplication of the coatings as necessary.

During the reinstatement a strong south-west wind developed whipping up considerable amounts of fine grit from the adjacent car park. This generated a form of air-abrasive blasting of such intensity that AMCS team members had to wear safety goggles and cover exposed faces and hands against the stinging blast. Although too weak to produce visible damage to the new coatings the grit stuck to tacky areas of new paint. Attendant HPR craftsmen confirmed this phenomenon as a regular occurrence at the church which may go some way to explain the increased decay on the sections of ironwork with a south-west aspect.

To present the church in good decorative order for the D-Day commemorations and to match the west ironworks' change in colour the remaining railings had, during the time of reinstatement, been overcoated with a gloss finish by a local paint contractor. Pre-treatment had been a brush down and bucket and sponge wash with fresh mains water.

MONITORING

Observations

The first two monitoring visits took place approximately six weeks and three and a half months after reinstatement.

Figure 9. Rapid paint deterioration as a result of continued underlying surface corrosion on unconserved railings. See also Colour Plate 20.

Thereafter visits were scheduled for every six months to coincide with the beginnings of spring and autumn.

July 1994: the visit was surprisingly informative, not for any noticeable change in the newly conserved west boundary ironwork but in the condition of the groups that had not been treated except for painting to the new colour scheme. Around the areas of scaly corrosion the fresh paint was beginning to blister and stain a rusty light brown colour.

September 1994: this visit showed a steady worsening of the paint on the unconserved railings (Figure 9). Meanwhile the first signs of deterioration had appeared on the test ironwork. On group 1 a bleached white bloom and roughened surface had appeared within some of the eyes at the centre of the areas cratered by corrosion. Group 3 had some discolouration of paint. The paint had darkened and produced a rainbow effect similar to oil on water both in the eyes of the eroded surfaces and where the stanchions entered the lead caulking. Despite the change in colour the paint film remained unbroken. Group 2 remained without any signs of deterioration. After inspection it was decided that the test ironwork should be redecorated to rectify the discrepancy of texture and matting of finish caused by the dust storm at the time of reinstatement. The affected areas were cleaned of grit and the whole of the west boundary ironwork was lightly abraded with fine grit-loaded Bear Tex scouring pads, sponge-washed with copious amounts of fresh water and redecorated with the gloss finish.

April 1995: this visit revealed on groups 1 and 2 a reappearance of the discolourations recorded the previous autumn. For the first time rusting was recorded on group 2: a small amount of rust had appeared where the paint had been chipped, which was most probably the result of traffic passing through the gateway. For the first time a change was noted at some of the inter-panel lap joints filled with mastic. The mastic remained pliable yet sufficiently stable to avoid sagging, however in a few cases it had disbonded from the metal leaving a hairline crack. Meanwhile the railings along the north, south and east boundaries had continued to corrode with resultant deterioration of the new gloss finish.

September 1995: the visit showed most of the ironwork to be in a similar condition to that recorded six months earlier. The notable exception was group 1 where the areas of white bloom had multiplied many times both in size and number. Careful scraping of an afflicted area revealed the discolouration to stem from the zinc substrate which had developed a friable white rusted surface.

April 1996: group 1 had now developed a white sheen over most of the areas eroded by previous corrosion, and paint was noticeably flaking from the south-west faces of palings (Figure 10). The condition of groups 2 and 3 appeared to have stabilized with no discernable increase in the areas of re-corrosion from the April 1995 visit.

Conclusions

A number of conclusions have already been reached despite the comparatively short monitoring period; these will be reconsidered as data continues to be collected.

Group 1

Currently this group appears to be the best protected against the re-corrosion. This conclusion may at first seem at odds with the visually deteriorated condition of the railings. However, close inspection show the deterioration is limited to the paint and zinc substrate.

Figure 10. White sheen developed on group 1, the zinc metal-sprayed group.

The protective mechanisms of thermal-sprayed zinc metal coatings are complex. The zinc coating is made from many tiny metallic particles rather than an impermeable coating and, if not adequately sealed, as appears the case in the church ironwork, moisture and oxygen will permeate into the matrix. The addition of aerobic moisture facilitates an electrochemical reaction between the two metals. While the reaction continues the iron will be protected from corrosion by the zinc which acts as a sacrificial anode in the corrosion cell. As corrosion occurs so the gaps between the zinc particles fill with zinc corrosion products until a barrier is formed. Thus the zinc will protect the iron both electrochemically and as a barrier against the environment.

The first appearance of zinc rusting was in the eyes of the craters and it is concluded that despite the rigorous cleaning regime some salts had remained to encourage moisture permeation through the paint and initiate corrosion of the zinc. The paint appears both dried and extremely thin, first impressions suggested that the regular blasting from the south west wind had contributed to the rapid deterioration. The effects of weather may be a contributing factor, however, as groups 2 and 3 do not appear as affected and the paint failure is attributed largely to the zinc corrosion. Once initiated the zinc corrosion would spread under the paint, weakening the bond with the metal causing the characteristic flaking now evident over most surfaces.

The objectives in painting the zinc were two-fold; firstly to seal the porous zinc and consequently slow down the rate at which the zinc would be exhausted by corrosion; secondly for decoration. It is very surprising the speed at which the paint has failed, particularly as the system had been employed on two earlier zinc metal-sprayed studio projects, one in an unpolluted rural environment and the other in an urban location, with long-lasting durability in both cases.

It is concluded that in the Royal Garrison Church application, the paint systems' impermeability was not sufficient to defeat the hygroscopic attraction of residual salts within the eyes of corrosion. It should be noted that of the three systems employed, System 1 had by far the lowest dry film thickness.

System 1 offers the most documentation to begin estimation of life expectancy. The *Atmospheric Corrosivity Values Chart*, issued by the Ministry of Agriculture, Food and Fisheries, indicates that the corrosion of zinc in the Royal Garrison Church locality is approximately 22g per m^2 per annum, equalling a annual corrosion of the railings zinc substrate thickness of 3.08μ. From this data the life of the now substantially unpainted, 100μ zinc coating would be calculated to be in the region of 32 years. However, this figure may prove to be optimistic as an alternative reference suggests a 15–25 year life span (American Welding Society Inc. 1974) which more closely matches the information contained within the BSI literature.

Group 2

This group has, so far, shown the least deterioration. The small amount of re-corrosion is the result of general wear and tear expected of functional exterior ironwork. It is still early to determine whether any deterioration has occurred below the coating, however, unlike groups 1 and 3, no discolouration can be found in the eyes of the corrosion. The thickness of the coating, when compared to the other systems, and the particular barrier properties of micaceous iron oxide, are likely to account for this, indicating that the system is performing to design.

Group 3
This group has proved to be the most difficult to draw conclusions from. The early discolouration of the paint, albeit in only a few places, led to an expectation of an early system failure overall. This has not occurred and the ironwork appears to have stabilized.

The early deterioration was particularly surprising as the two primers are specifically designed for application over rusty metal surfaces. The areas of deterioration are within the corrosion eyes and around the stanchions where they enter the stone coping. It is suspected that, as with group 1, residual salts within the corrosion eyes have initiated re-corrosion. However, the paint film does not appear in any way broken, and rusty solutions have not emerged. It is concluded that in the contaminated eyes the paint has disbonded from the metal, possibly as a result of a brief initial reaction between the drying primers and salts, but the paint barrier has not been breached and corrosion has now subsided. The deterioration around stanchions is attributed, despite the thorough making good, to burning of the paint during pouring of the hot lead fixing. The paint was noticeably affected by the procedure and blistered at least 15mm upward from the final top surface of the lead

Mastic
With the exception of the few hairline cracks, first noticed during the April 1995 visit, the material has performed well in respect of maintaining shape and elasticity. The reasons for the limited fracturing at the interface with the metalwork is difficult to determine. It appears to be poor adhesion between mastic and paint in those places rather than the cracking sometimes experienced in rigid fillers through shrinkage or movement of materials. In the longer term the fracturing raises the question of water penetration and retention through capillary action.

SUMMARY AND DISCUSSION

The severe corrosion of the Royal Garrison Church metalwork and the harsh, local environment to which it is exposed, made it a valuable yet challenging subject for practical research. While the cleaning and coating programme, broadly based on BSI recommendations, was comprehensive, restoration of the metal fabric was restricted to the few localized consolidation repairs needed to return structural stability. The methods and coating materials used were chosen for the regularity with which they appear in typical specifications. The use of modern welding equipment, although not always appropriate to the repair of historic ironwork, was in this case the instrument of minimum intervention. The MAG welds, once blended into the surrounding iron and over-painted, were unobtrusive, but they should become obvious to the practised eye during the course of any future conservation.

Test papers of Potassium Hexacyanoferrate (III), usually referred to as Potassium Ferricyanide, are by far the most commonly cited method of detecting salts. The procedure is easy to carry out and reasonably effective. However, a question is raised by the failure of the papers to detect salts where later corrosion appeared in some of the corrosion eyes. This may be the result of salts being harboured within the fibrous wrought iron thus avoiding contact with the paper. Curiously, given the papers' universal recommendation, it was not possible to locate a commercial supply, perhaps suggesting a lack of demand through infrequent practical application. The value of differentiating between chlorides and other contaminants is debatable for a project such as this.

Cleaning

To date, practitioners cleaning historic ironwork have largely been divided into two camps. Traditionalists have employed long-established, comparatively low-technology methods of paint and corrosion removal, for example hand or flame cleaning. The other approach has been to use the various types of air-abrasive blasting primarily developed for modern industry. The latter method now predominates. The domination of air-abrasive cleaning is in part financial. Over large areas the cost of capital plant and compressed air generation is offset by the speed of cleaning, although cleaning rates can be dramatically reduced when working on heavy corrosion and three-dimensional objects. The other important factor in the rise of air-abrasive cleaning is the development of modern, technically advanced coating systems which require the metal's surface to be clean and abraded to an agreed level to guarantee the paint's maximum performance. The BSI standards and paint manufacturers' literature point to air-abrasive cleaning as the most effective method of surface preparation. Against this background metal conservators are now questioning the suitability of rigorous cleaning, and indeed recleaning, of historic metals and are searching for less invasive treatments.

Flame cleaning has begun to regain some of its lost popularity. The technique can be very effective in removing both old coatings and corrosion, and, if carried out with care, will not damage the metal's surface. It should be noted that flame cleaning will disbond millscale, may buckle thin sections and particular caution should be exercised around fixings which may be affected by excessive heat. For cleaning a small area, particularly on site, the cost of flame cleaning is likely to be less than air-abrasive blasting. Interestingly, for the Royal Garrison Church ironwork, flame cleaning proved an effective pre-treatment to blasting, removing the bulk corrosion product which tended both to slow the rate of cleaning and, if not scrupulously filtered from the grit, block the blast delivery nozzle. Surprisingly, it is generally the case in conservation specifications where flame cleaning is stipulated that no reference is made to the relevant section in BS 7079.

Salts

Both dry-abrasive and flame cleaning proved to be successful in removing paint and corrosion. Yet neither method tackled the problem of salts contamination, and in both cases the iron had to be washed to remove salts. These were very prolonged processes which left the metal's surface rusty, and ultimately appear to have been insufficient preparation for at least two of the coating systems. Wet air-abrasive cleaning did go some way in tackling the salts contamination. However, further water washing was required, and the surface was left both abraded and slightly rusted. High-pressure water blasting was included in the project following positive reports from contractors who use the system to clean marine installations, and from private conservators who used a similar higher-pressure system to remove marine-borne salts from zinc sculpture.

During the Royal Garrison Church trials, high-pressure water blasting was very successful in removing salts, corrosion and decayed paint to leave a clean metal surface. Gingering did of course occur. The system also had the advantage of preventing the dusts associated with dry abrasive blasting. An anticipated limiting factor not encountered on the sturdy church ironwork is that delicate pieces, for example decorative leafwork, may distort under the high-pressure jet. Unfortunately the difficulty in operating the equipment, and time constraints on developing the equipment for easier use, resulted in the system falling short of the project criterion. Looking to the future the system obviously has great potential and it is considered that a reduction in the water volume, while maintaining the high pressures, may solve the difficulties encountered in directing the jet. Another development that would assist conservators using the equipment, particularly working on objects with delicate components, would be to have the pressure regulator and gauge fitted to the jetting gun rather than the pump. The interest in and development of pressure watercleaning is considerable, both from industry and conservators and it appears certain to enter wider use.

Coatings

Given the nature of the UK climate, the main function of coatings for exterior ironwork is protective rather than decorative. Science has been harnessed to develop modern coatings that are both extremely effective in protecting new metal surfaces and are long-lasting. However, as with cleaning, there is a now a need for a reappraisal of the appropriateness of these coatings for historic exterior ironwork. Such a review should seek to address issues such as the perceived difficulty in removing certain 'high tech' coatings, difficulties of on-site maintenance, concerns over the environment and health and safety and the recognition that many of the currently used coatings may not be suited for application to surfaces prepared by the less intrusive cleaning systems.

A final judgement of the three selected coating systems is still some time away. Certainly System 1 has not performed as expected, although it is now concluded that the paint system was, while conforming with the manufacturer's specifications, in this case inappropriate. It would be interesting to discover whether marine-corroded steel also retains salts tenaciously after cleaning sufficient to corrode newly applied thermal sprayed zinc. Meanwhile System 3 appears to have been affected by residual salts and by the use of traditional hot lead to caulk fixings. This far into the monitoring System 2 appears to offer the best combination of protection, aesthetics and cost-effectiveness. However, this last point may be diminished by time as BS 5493 indicates that the system will need far more maintenance than System 1 in the long term.

Looking to the future, there a strong movement within the conservation field to return to more traditional paint types, which were considered to fall outside the scope of this project. Paint manufacturers will also continue to respond to the demand for environmentally-friendly products; this progression can only be beneficial to those charged with a conservative approach to metalwork repair especially with due regard to site maintenance.

The inclusion of a filler into the long-term test programme was driven by regular inquiries to the AMSC from specifiers seeking an alternative to red lead putty. Densoseal 16A contained many of the attributes required of a sealer, yet some of the positive features may limit the range of applications. For example, the springy dough-like consistency made it hard to shape the material precisely to the shape of the adjacent iron. This can also be the case with red lead depending on the viscosity of the putty. The slight disbondment of 16A with painted surfaces also detracted from the generally good impression. Hard-setting polyesters and epoxies are sometimes used to fill water traps but their limited flexibility raises yet again the possibility of cracking.

An alternative approach to this is to consider relying less on fillers and more on adequate drainage of water traps. Poorly sited or cracked fillers may retain moisture in crevices, or, in the worst cases, dam back water which would have otherwise flowed away. However, the clearance of accumulated debris from unfilled water traps requires regular maintenance. It is the lack of provision for this very maintenance which in itself is responsible for the advanced state of deterioration of much of the exterior ironwork in the UK. And it is the continuing failure to recognise the importance of maintenance that has required the transfer, sometimes inappropriately, of methods and materials developed primarily for new, industrial works in a search for longer and longer 'years to first maintenance'.

BIBLIOGRAPHY

J. Ashurst and N. Ashurst, Practical Building Conservation Series, Volume 4: *Metals* (Aldershot: Gower Technical Press, 1988).

British Standards Institution, BS 5493: *Code of practice for the protective coating of iron and steel structures against corrosion*, (London: British Standards Institution, 1977).

British Standards Institution, BS 7079: *Preparation of steel substrate before application of paints and related products* (London: British Standards Institution, 1989–94).

British Standards Institution, BS EN 22063: *Metallic and other inorganic coatings. Thermal spraying. Zinc, aluminium and their alloys* (London: British Standards Institution, 1994).

British Standards Institution, BS 3900: *Methods of test for paints* (London: British Standards Institution, 1994).

I. P. Haigh (ed), CIRIA Report 93: *Painting steelwork* (London: CIRIA, 1993)

Ministry of Agriculture, Fisheries and Food, *UK atmospheric corrosivity values* (London: MAFF Publications, 1986).

L. L. Shreir, R. A. Jarman and G. T. Burstein (eds), *Corrosion* (2 vols) (London: Butterworth Heinemann, 1994), 13:46.

American Welding Society Inc. *Corrosion tests of flame sprayed coated steel – 19 year report.* 1974

ADDRESSES

Acrylon Environmental Ltd, PO Box 684, Amersham HP6 6DX; tel: + 01494 726890, fax: + 01494 726890

Akzo Coatings plc, 135 Milton Park, Abingdon, Oxon OX14 4SB; tel: + 01235 862226, fax: + 01235 862236

British Flow Plant Group Ltd, 7 Thames Road, Barking, Essex IG11 0HN; tel: + 0181 591 7799, fax: + 0181 594 4195

Elcometer Instruments Ltd, Edge Lane, Droylsden, Manchester M35 6BU; tel: + 0161 371 6000, fax: + 0161 370 4999

English Heritage Ancient Monuments Laboratory, 23 Savile Row, London W1X 1AB; tel: + 0171 973 3000, fax: + 0171 973 3001. The AML has a web site containing information and project news: www.eng-h.gov.uk/. Full text of metallographic examination is available in AMLAB report 35/95.

English Heritage Picture Conservation Studio, Inner Circle, Regents Park, London NW1 4PA; tel: + 0171 935 3480, fax: + 0171 935 6411

Hodge Clemco Ltd, Orgreave Drive, Sheffield, Yorkshire S13 9NR; tel: + 01142 540600, fax: +01142 540250

The Ironbridge Gorge Museum, Blists Hill, Ironbridge, Telford, Shropshire TF7 5CU; tel: + 01952 583003, fax: + 01952 588016

Merck Ltd, Hunter Boulevard, Magna Park, Lutterworth, Leicestershire LE17 4XN; tel: + 0800 22 33 44, fax: + 01455 55 85 86

Metallisation Ltd, Pear Tree Lane, Dudley, West Midlands DY2 0XH; tel: + 01384 252464, fax: + 01384 237196

Power Blast International Ltd, Southern Trade Centre, Camberley, Surrey GU34 2QG; tel: + 01420 588450, fax: + 01420 588426

RS Components Ltd, PO Box 99, Corby, Northants NN17 9RS; tel: + 01536 201234, fax: + 01536 405678

Walden Precision Apparatus Ltd, The Old Station, Linton, Cambridge CB1 6NW; tel: + 01223 892688, fax: + 01223 894118

Winn & Coales (Denso) Ltd, Denso House, Chapel Road, London SE27 0TR; tel: + 0181 670 7511, fax: + 0181 761 2456

ACKNOWLEDGEMENTS

In addition to their English Heritage colleagues Fred Powell, Helen Hughes and David Starley, the Architectural Metals Conservation Studio team would like to thank Barry Knight of the Ancient Monuments Laboratory for his help and support during the Royal Garrison Church project, and during earlier projects to detect and treat the corrosion of architectural ironwork.

The team would also like to thank Rupert Harris, private conservator, for his help in identifying suppliers of the high-pressure water jetting equipment and waterborne coatings used in the project.

AUTHOR BIOGRAPHIES

Keith Blackney is a private conservator specializing in the conservation of architectural metalwork. He has worked widely in mechanical engineering, firstly in the development of construction plant and railway vehicles, and later in the prototyping of special purpose machine tools. He is a former member of the Architectural Conservation team at English Heritage, working at the Architectural Metalwork Conservation Studio from 1988–97. His interests are in the adaptation and development of methods from other conservation disciplines and from industry for the non-destructive treatment of exterior exposed metals.

Bill Martin trained as a stone conservator and ran his own studio for nine years before moving to the Council for the Care of Churches where he was Conservation Officer. He joined the Architectural Conservation team at English Heritage in 1989, and his responsibilities include project management of research into the decay and conservation of historic tile pavements and the evaluation of masonry consolidants. He also managed the English Heritage Architectural Metals Conservation Studio in Regent's Park, London until its closure in 1997 and coordinates technical advisory work for the Architectural Conservation Team.

Annex

English Heritage's research programme: a schedule of projects on historic building materials decay and their treatment 1992/3-1997/8

AC1 Smeaton Project on historic mortars durability

An investigation into the effects of set additives on lime-based mortars: Phase I in association with the International Centre for the Study of the Preservation and the Restoration of Cultural Property (ICCROM), Rome and the Department of Conservation Sciences, Bournemouth University.

AC2 Masonry consolidants

The long-term field assessment of the behaviour of *Brethane* alkoxysilane consolidant trials on stone monuments and laboratory-based methods to locate and evaluate residual consolidant remains in weathered stone.

AC3 Floor wear

The field assessment of pedestrian traffic wear on historic stone pavements. The development of management and protection regimes to ameliorate potential damage. The field assessment of decay, wear and damage of medieval encaustic tile pavements in cathedrals. The development of management and protection regimes for their preservation.

AC4 Tile pavement decay

The long-term field monitoring and assessment of exposed medieval tile pavements in situ on open ruined monuments and archaeological sites. The development of management and protection regimes that safeguard their welfare without damaging the archaeological integrity of the sites.

AC5 Polishable limestone decay

In situ environmental monitoring, decay mapping and laboratory testing to understand the decay processes and rates of erosion of Purbeck stone and other decorative polished limestones. Publication of a guide to polishable limestones in the UK. In association with the Surveyors of the Fabric of Chichester, Rochester, Lincoln, Salisbury and Norwich cathedrals.

AC6 Sandstone decay

A literature review and preliminary evaluation of decay systems in UK sandstones to define and explain observed deterioration phenomena. The development of a simplified decay mapping system to measure rates of decay.

AC7 Anti-graffiti barriers & their removal

An assessment of the effectiveness of wax-based and other anti-graffiti barriers, and their periodic removal after paint attacks, to determine whether they contribute favourably to the long-term protection and welfare of friable historic masonry.

AC8 Structural fire protection

The establishment of the standard performance in fire of historic panelled timber doors and systems of upgrading. The preparation of a fire engineering approach to fire safety in historic buildings. A review of research into the performance of structural cast iron in fire.

AC9 Lime and lime treatments

The characterization of selected building limes in the UK, including both domestic and foreign non-hydraulic and hydraulic limes. The production of a Directory of Building Limes. An assessment of lime treatments (lime repairs and lime shelter coating) and their effectiveness.

AC10 Underside lead corrosion

The long-term determination of all parameters affecting underside lead sheet corrosion and the development of preventive and remedial treatments that do not materially affect the special architectural interest of historic roofs. In association with the Lead Sheet Association, the Historic Royal Palaces Agency, Liverpool John Moores University, the National Trust and Bristol University.

AC11 Timber decay & moisture ingress

An assessment of structural timber decay and its treatment in English cathedrals: a review of current procedure and the development of improvements in practice.

AC11/2 Woodcare

A study of the inter-relationships between the environment, the fungus *Donkioporia expansa* and deathwatch beetles (*Xestobium rufovillosum*) in cathedral roof spaces with particular attention to moisture ingress, the ageing of timbers and natural predation. In association with the European Commission Directorate General XII, Ridout Associates, Birkbeck College, London, The Jodrell Laboratory of the Royal Botanic Gardens, Kew, University College, Dublin and TNO-Bouw, of Delft in the Netherlands.

AC12 Masonry cleaning

A review of currently available masonry cleaning systems in the United Kingdom and the development of policies and guidelines towards them. The redrafting of the British Standard Code of Practice *Cleaning and Surface Repair of Buildings* BS 6270 Pt.I: 1982. In association with the British Standards Institution and its Subcommittee B/209/7.

AC13 Fire safety in cathedrals

A study of the fire safety provisions in cathedrals in the light of the recommendations in the Bailey Inquiry Report. The development of joint guidelines with the Cathedrals Fabric Commission for England.

AC14 Terracotta decay and conservation

The development of a lexicon of terminology for architectural ceramics. An international literature review of research on the manufacture, decay and treatment of architectural ceramics. Studies of the special sensitivities of architectural terracotta to soiling, decay and the impact of cleaning.

AC17 Stained glass

A study of the factors affecting the durability of lead cames in stained glass. A literature review and general search of industrial data in relation to the modern use of butyl mastics. A short study of polycarbonate sheet protection for stained glass with respect to its decay in UV light, cost-in-use and architectural impact in historic settings. A literature review regarding the use of Paraloid B72 as a surface consolidant for stained glass.

AC19 Cathodic protection

The development of a live cathodic protection system, by technology transfer, for corroding isolated buried metal masonry cramps in historic buildings. Refinements to non-destructive surveying techniques and key-hole surgery to limit collateral damage during the installation of the system.

AC20 Stone slate roofing

An investigation into the state and causes of the decline in the stone slate roofing industry in the Peak District National Park and the development of ways and means to counter the situation. In association with the Peak District National Park Joint Planning Board, Derbyshire County Council and the Construction Sponsorship Directorate of the Department of the Environment.

AC21 Mosaic clad concrete

An investigation into the causes of decay and deterioration in ceramic and glass mosaic-clad concrete on twentieth-century buildings.

AC23 Hydraulic mortars & grouts

Comparative studies of hydraulic binders for conservation mortars, plasters and renders in the UK. Confirmation of policy towards repair recipes for use in specifications.

AC24 National mortar, sand & aggregates library

The development of a national reference collection of sands and aggregates for building conservation mortars. Characterization of samples and the publication of a national directory.

AC27 Earthen structures

The development and promotion of research and other technical studies in the decay and conservation of earthen architecture in the UK.

NOTES

A new five-year forward programme of research is to be agreed, programmed and resourced before the end of May 1998 and will be published in a future volume of the *Transactions*.